Breakthroughs

# BREAKTHROUGHS

## Realizing Our Potentials Through
## Dynamic Tricky Mixes

# KEITH NELSON

NEW YORK

LONDON • NASHVILLE • MELBOURNE • VANCOUVER

# Breakthroughs

## Realizing Our Potentials Through Dynamic Tricky Mixes

Published in New York, New York, by Morgan James Publishing. Morgan James is a trademark of Morgan James, LLC. www.MorganJamesPublishing.com

Proudly distributed by Ingram Publisher Services.

ISBN 9781631956690 paperback
ISBN 9781631956706 ebook
Library of Congress Control Number: 2021911193

**Cover Design by:**
Rachel Lopez
www.r2cdesign.com

**Interior Design by:**
Christopher Kirk
www.GFSstudio.com

Morgan James is a proud partner of Habitat for Humanity Peninsula and Greater Williamsburg. Partners in building since 2006.

Get involved today! Visit MorganJamesPublishing.com/giving-back

*To your future, to worldwide collaboration and cooperation,*
*and to the future of this precious Earth.*

# TABLE OF CONTENTS

# ACKNOWLEDGMENTS

First, to my incredible wife, Kathrine Nelson. Thank you, Kathy, from the bottom of my heart for all of your love and support, all you do, and all that we have shared.

A salute to the wooden "octopus" in a Pennsylvania forest, which helped bring Kathy and me together.

Also, deep appreciation to my many wonderful colleagues in research and brainstorming endeavors—and to the fun of it all.

Hats off to the Maine Media Workshops. One workshop on photography and another on writing poetry were seminal influences on widening the scope of my creative endeavors. Further hats off to my central mentors at Yale, where I was a graduate student.

Thanks also to David Hancock and his wonderful team at Morgan James Publishing for accepting my work and for their excellent and cheerful assistance at all phases of manuscript preparation, publication, and marketing.

Finally, to my dozens of graduate students and thousands of undergraduate students over the years—you taught me so well and opened many doors in my mind and spirit.

# PREFACE

This is a book about *breakthrough* advances and how and why they happen. Many are from my own experiences as a scientist, and many others come from my own worldly adventures. Moreover, key advances reported by non-scientists and scientists all over the world are also part of the excitement. I have chosen to include breakthroughs of all kinds—skill acquisitions, inventions and innovations, high-performance episodes for skills already in our repertoire, recovery of lost abilities, new significant connections between people and between people and nature, and more. The breakthroughs covered are not in any way meant to be a historical "greatest moments" listing, but rather, an accessible, provocative, and inspiring assortment that you and most readers can take in and relate to situations in your own lives and communities.

My first hope is that you will just find these breakthroughs fun to explore.

Beyond that, I also hope that you will ask yourself the following questions about each episode, each advance: Why didn't I know this breakthrough was even possible? What can I learn from this breakthrough? How does learning about different problem areas and issues help in understanding complex dynamics utilizing creative thinking toward powerful strategies and new pathways to breakthroughs? Where else may I apply a similar "tricky re-mixing" of conditions to help myself and others I care about achieve new breakthroughs?

Dedicated to your future, to worldwide collaboration and cooperation, and the future of this precious Earth.

**Keith Nelson,**
Eagle Spirits Farm, Eagle Spirits Creative Breakthroughs,
and Pennsylvania State University

# INTRODUCTION

**D**ynamic tricky mixes, huh? Yep, the book is centered on learning how to work with the dynamics of what is really happening in situations and moving beyond simplistic expectations and strategies. The incredibly wide set of examples of breakthrough advances that you encounter puts you in the front row seat for discovering the fascinating tricky moves that create remarkable advances in people's lives.

Truly understanding complex goals and events we care about deeply is difficult and often frustrating. Yet, whatever we decide to do will spin out in a positive or negative direction, subject to inescapable complex dynamic forces. Too often, even the most experienced, the experts in the field, will fall back on overly simplistic strategies for change. With confidence on their side, they fall into a deep trap, pouring generous resources and time into narrow procedures and practices, while starving resources for valid monitoring of what is really happening versus what had been so confidently hoped for.

Dynamic mixes always unfold in real time and in complex and somewhat unpredictable ways. But despite these dynamic complexities, we can approach any project with savvy awareness. Further, we can add deliberate dedication to experiment with conditions and make detailed and honest observations on whether our project is going downhill, going uphill slowly, or racing along just as we had hoped. We can become flexible "tricky mixers." We can become tricky mixers who are fabulous at working with collaborators with varied backgrounds and skills but who approach formulation, monitoring, experimentation, and revisions of plans and procedures with an active tricky mix perspective.

Now, we will take one sneak peek at a later assortment of situations where resources were poured in with high expectations of success, yet where we there was persistent failure across a great expanse of time; then, for the same situations, discover later truly shocking advances. We will see stark contrast between unsuccessful attempts at tricky mixes and positive and dynamic tricky mix success.

## Wasted Attempts

From age two to nine, Ted encountered many mixes, many attempts to trigger social skills and language. Among the woefully inadequate strategy mixes were several forms of behavioral treatments and a computer interaction to build better engagement—first with the computer, then for engaging with other human beings. Neither language skills nor social skills were advancing.

## Contrasting Positive Tricky Mixes

Some therapists and teachers would be inclined to believe that such a kid—one on the Autism Spectrum—would be carrying such a dysfunctional brain that any novel attempt at language therapy would also fail. We will see that a radically new tricky mix of conditions transformed what this child was capable of over the next six years. Feel free to speculate about what forms or styles of new language/social interactions could trigger rapid language advances for Ted.

Welcome to the world of *Breakthroughs,* achieved through dynamic tricky mixes. You will be guided back and forth around seemingly unconnected areas while learning the information. Who else claims we can find core similarities in this partial sample of events? Helping a crippled old dog. Unleashing art potential in a five-year-old child. Getting a corporation unstuck and moving boldly into innovation. Confronting "Draculas." Folks with outrageously good memories. Shooting "cows." Bringing language and reading into an autistic child's life. Not becoming a mountain lion's dinner. Making psychotherapy effective. A safer environment. School reform. Snowballs. Rehab after stroke or injury. Scrapping long-standing, expensive plans that are disasters but had been resistive to evaluation. Reconnecting to nature. Becoming a quadra-lingual child. Octopus design.

Ending misguided, highly destructive environmental exploitation public policies. Making sensible action choices in a viral pandemic.

This book is timely and essential because one of the most deeply ingrained tendencies in our modern thinking and social policy is the reduction of complex phenomena into simplistic "silver bullet" strategies and conclusions. A key antidote is the introduction of *positive dynamic tricky mixes*. Such mixes work within the inherent complexities of situations to dramatically trigger unmistakable advances.

*Breakthroughs* guides us through a broad "banquet" of remarkably diverse events and contexts. They all serve as cautionary tales against simplistic thinking in education, science, environmental issues, social policy, economic policy, entrepreneurship, and interactions among contrasting cultures. Moreover, it becomes evident that the same basic dynamic tricky mix framework invites and inspires new paths to desired changes in our personal lives. We can learn to make truly significant changes in our lives, and we can further learn how to coordinate and synergize progress whereby we join with a handful of others or with a crowd of people.

Each chapter will introduce you to amazing events that emerge out of very complex conditions. These events cause jaw-dropping fascination, reason enough to race through all the realms and all the twists on how tricky mixes can be so powerful. But as you absorb the full panoply, several integrative conclusions emerge.

First, the events help to loosen the frameworks of our thinking to allow patience and serious attention to the ways in which real-world "mixes" of multiple, complex, interacting conditions lie behind the outcomes we care about. Again, learn to fight off the allure of anyone hawking a deceptive, simple, seemingly easy route.

Second, they inspire hope and confidence that new attempts at solving complex problems can provide genuine innovation and enhanced effectiveness. You will gain confidence that new dynamic approaches may jolt free what had appeared to be hopeless, fully stuck situations.

Third, these events—both singly and collectively—suggest strategies for protecting ourselves from unplanned/unexpected toxic consequences and also from the waste of resources and time and human capital that flows from simplistic thinking.

Fourth, we gain insight into how to give up the "bad habits" of familiar, easy to remember, emotionally-invested, but fatally flawed simple approaches.

Fifth, a review of how dynamic tricky mixes have already been highly effective across these amazing and complex events generates strategies for designing, implementing, and evaluating new Mixes of conditions that have never been tried before. Such innovations, such new mixes and re-mixes, are crucial for both individual and group progress. Expect to learn how dynamic systems thinking draws patient attention to what conditions matter for particular kinds of advances/breakthroughs, and even more crucially how the patterns of the conditions converging together (or not converging) are at the heart of how rapidly advances can occur.

Sixth, we all gain a deep appreciation of the importance of finding causal experiments that illuminate—despite the high complexity of events—where our investments of time, money, and hope actually pay off.

Seventh, as noted earlier, dynamic mixes always unfold in real-time and in complex and somewhat unpredictable ways. But we can approach any project with savvy awareness and a dedication to intensely experiment with conditions. Further, we can make detailed and honest observations of whether our project is going downhill, going uphill slowly, or racing along just as we had hoped. We can become flexible tricky mixers who know from experience that these kinds of active innovation-and-adjustment processes are do-able and well worth the effort. We can become tricky mixers who are fabulous at working with collaborators with varied backgrounds and skills but who all approach formulation, monitoring, and revisions of plans and procedures with an active tricky mix perspective.

In a nutshell, you will discover refreshing and powerful new mixes of conditions, which you can launch for more effectiveness with your efforts in almost any endeavor. Wherever you live, whatever your identity and self-concept, and whatever your key projects, by plunging into the events and episodes of *breakthroughs,* you will find numerous relevant aspects. This holds true, in part, because some of the breakthroughs we discuss are in areas of life that every person encounters; they will immediately seem familiar. But beyond that, the wild range of examples beyond your own experience will prove relevant because they illustrate—and make real—the same tricky mix dynamics that feed into whatever projects you have already undertaken and whatever you may choose to launch or join in the future.

So, please, prepare yourself for a wild ride across widely different problem areas and the discovery of *Complex Dynamic Processes* for all of them. You will find that similar *dynamic tricky mix* innovative thinking facilitates *breakthroughs* in all areas.

Chapter 1

# FROM SNOW PATTERNS TO HUMAN PATTERNS: THE ESSENCE OF DYNAMIC TRICKY MIXES

As I write these words, I enjoy sitting in my study, high up in a remodeled old farmhouse at our Pennsylvania farm, "Eagle Spirits Farm." It is early March 2021, and we have already experienced several snowfalls and snow melts. When I glance outside, I view meadows still partially covered with snow.

Below, I will include some snow-related photographs, which reveal the dynamic processes that are central to this book.

First, though, reflect on how difficult snow predictions and other weather details continue to be for scientists. Why, in this day and age, despite extensive technologies worldwide—technology that continually monitor temperature, humidity, winds, clouds, and more—and despite fast computer matrices and thousands of trained weather scientists, can the models not tell me more precisely when the meadows here at the farm will get snow and how much?

The answer lies partly in the highly complex, highly dynamic, and rapidly changing *mix* of conditions that feed into current weather any place on earth and then feed future weather patterns. *Dynamic systems* are at work, for sure.

We emphasize the *trickiness* that emerges, not only for weather predictions but for patterns of human planning and action. It becomes *tricky* to discover an accurate account of what the current *mix of conditions and their dynamic interplay* are and whether they are leading into new snowfall on our meadows in the near future. So, both understanding the current dynamic patterns and predicting future patterns turn out to be *prone-to-error, complex,* and *elusive*—in short, *tricky.* The same cautions apply to the attempts we make to create better action plans, procedures, collaborations for teaching children, rehab strategies for adults, the composition of art, music, and dance choreography, global warming and pollution controls, reductions in inter-group strife, and all other domains.

Sometimes, the particular *dynamic tricky mixes* we launch prove to have negative effects rather than the positive outcomes we hope for or expect. However, by maintaining a continual awareness of the complexity and trickiness of dynamic processes, we place ourselves in much better positions to make new analyses and revised plans, to move beyond failures to new, highly successful *dynamic positive creative mixes.* The many discussions, events, and phenomena within this book are intended to lead you and others to increased awareness, the discovery of a rich new set of strategies and experimental moves, and exciting new progress on your agendas in any domain.

Now, consider a maxim that sums up much of what you will encounter and can apply in your own life.

> Dynamic tricky mix thinking calls attention to greater complexities in the dynamics of situations than what is usually considered. In turn, given such heightened awareness and understanding of the dynamics in place, there are greatly-enhanced possibilities for individuals and groups to create highly engaging and highly successful new pathways. "Tricky" new "mixes" of strategies, partners, social connectedness, careful and valid monitoring of progress, and dynamic adjustments, or "re-mixes," emerge and dramatically transform the rate of progress toward key goals.

Next, let's return to the natural world and the dynamic patterns concerning snow.

In this spirit, explore the photographs of snow below. Take note of what you see in them, and then read my commentary following them.

*Melting Snow Patterns, Puzzle One—What Do You See?*

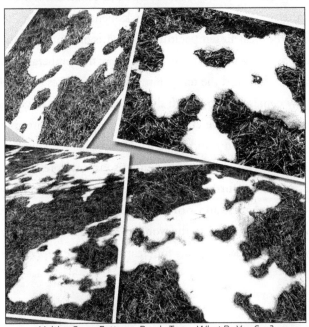

*Melting Snow Patterns, Puzzle Two—What Do You See?*

If you saw more than "just snow" in these pictures, you were using a cognitive ability that has shown its power repeatedly, across nearly two million years. It's the ability to re-interpret what we see, assessing it as more than the usual, familiar, literal object or event before us in a scene. In every sense, there were no ghosts or dragons or dogs in the snow scenes I just shared from our local environment, our Eagle Spirits Farm. You own whatever creatures your mind generated in response to these shots of simple, everyday snow that is in the process of slowly melting away.

A more simplistic view is that our land here was covered with two inches of snow, which gradually and smoothly declined to almost nothing, then to no snow at all.

Complex dynamic systems, however, were in play all through the week. First, when the snow fell, it did not deposit evenly. Instead, for each patch of ground, the amount received was influenced by a *tricky mix* of local air temperatures, wind shifts, and uneven ground surfaces, including the presence of rocks and sticks and tufts of grass. Similarly, each day after the snowfall, the speed and pattern of melting snow led to the *emergence of particular patterns in the newly exposed ground and in the remaining snow*. These would have been exceedingly tricky to predict before their occurrence. Among the *tricky mix of dynamic conditions converging* would be at least these: the size and warmth of the rocks beneath the snow, the warmth of the bare ground, the local air temperature, any swirls of wind, the angles of the sun and the clouds obscuring the sun at times, and how the temperature of the ground (already revealed) influenced the temperatures across patches of remaining snow.

As the snow comes and goes with dynamic complexity, so goes all aspects of our individual and cooperative efforts. Further, so goes the interplay of *complex tricky mix conditions* in nature and human relationships with nature.

You are invited to go beyond any comparisons, interpretations, and conclusions that I draw in these further chapters, ones that traverse seemingly dissimilar experiences, events, players, and situations. Please generate and carry into your future endeavors new insights into the wonders and frustrations of highly complex *dynamic systems* and new action plans for creating positive *tricky mix* solutions.

# Chapter 2

# ASSORTED EYE-OPENERS

Imagine a better future. What would you like to see enhanced in your own life and society and the world at large?

The processes for creating change are strikingly similar, regardless of the topic or whether one, a few, or millions of individuals are affected. In all cases, too-simplistic approaches, overconfidence, premature conclusions, and "sacred cows" must be avoided.

## Consider our Family Dog Sail

Sail was a spirited, chunky, lively golden Labrador retriever. She was lucky to live on a farm and chase rabbits and groundhogs, relishing long walks in the woods. Sail personified strength and energy, bulldozing through snow or bushes or dirt or any other obstacle to get where she wanted to go. This beautiful and muscular family member loved our farm and her active life.

Then, quite suddenly but with age, her agility and strength seemed to disappear. Sail could no longer hop into our car for a ride or even walk up the steps to the farmhouse porch. It seemed for a while that her active life was over and that we might very well end up hauling her around in a trailer behind a small tractor so she could visit the ponds or the woods.

One simple—and ultimately incorrect explanation—was that her joints and muscles were just too worn out to ever function well again. One dead-end treatment was the provision of the single aging-dog supplement, glucosamine. This made her lose a little weight, but her weight was not the critical variable.

It turned out that a more complex solution, a more complex "mix" of components, made all the difference. Once we started her on this new mix—just a few days later and for years afterward—Sail could again climb the stairs, make the long hikes, and chase the ever groundhog she wanted. She was not hopelessly physically damaged nor did she need surgery of any kind; that was proved by her rapid recovery.

*Reader alert.* Are you able to guess what new *mix* brought Sail into recovery?

The right "tricky mix" for our beloved lab was borrowed straight from college campuses and rehab clinics, where strategies for athletes with sprains and tears of muscles and ligaments and tendons reign: Sports Complex pills consisting of relatively high doses of three components rather than just one component, namely glucosamine plus MSM plus chondroitin. Since another one of Sail's traits was to gobble just about anything thrown her way, it was child's play to get her to start taking these pills.

Drum roll—*bam, ba bam, ba bam, bam, bam!*

Three days later, our old Sail was back—climbing stairs, hopping into the car, frolicking on hikes, and eagerly chasing rabbits. And as she became more active week by week, that exercise further fed into her strength and recovery program. What a relief!

This almost miraculous recovery demonstrated several aspects of "dynamic mixes" as a route to new progress. For Sail, her recovery proved that she was not too worn down physically and not suffering from some degenerative illness. Her body was still "ready" for an active, engaged life, but this readiness was not evident until a new combination of conditions was dynamically mixed with her ongoing eating and exercise patterns and her physical, physiological makeup.

*Our dogs, Skye, Banjo, and Sail, each of which got stuck—aged and lost mobility—and then recovered quickly with a Dynamic Re-Mix of nutrition.*

## Illusions of Complete Analysis and Perfect Planning

Other instances of dynamic mix processes highlight the fact that even when extraordinary efforts have been maintained for long periods, an effective mix is not guaranteed to emerge. I was in a Minnesota school, hoping that my research lab's efforts to create a new mix of software and teacher strategies would pay off in dramatic gains for poor readers in elementary school when the Challenger space mission blew up soon after liftoff. Much more to come about children's learning . . .

For now, consider the Challenger disaster as a case study in the complexity of mixes when large teams of skilled professionals work toward a fail-free technological accomplishment. All of the Challenger spaceship's systems and all of the required rocket technology that boosted it into a high orbit were checked and rechecked with multiple backups and alerts. Nevertheless, it blew up.

Large and numerous committees found the whole disaster puzzling. Failure should have been precluded by the extensive planning and backups, the fail-safes. Into this morass steps the physicist Richard Feynman. He, as an outsider to the project attending a dinner event that brought experts together on the matter, used a rubber band and a glass of ice water to demonstrate that the likely critical variable was temperature and its effects on the flexibility and density of the rubber seals between rocket stages. The whole orderly yet complex sequence became a deadly *negative* tricky mix when the temperature in Florida at launch time was significantly colder than any testing and simulations had covered. The seal failed and fed the dynamic mixtures for a catastrophic explosion. As an outsider to the whole project, Feynman was an active and creative thinker and had quickly ferreted out the relevant causative *mix* behind the surprising disaster.

## Breaking into Words from the World of Autism

We can see in examples of children's learning the same dynamic processes at work. Some shifts in learning incorporate radical, surprising changes in context. This proved true for one boy with autism, Tommy.

He had been given countless behavioral trials calling attention to a single word, such as "cookie." Any correct attempt at using the word would receive an immediate reward in this procedure, whether stars or candy or other items. Years of training did not produce results: this boy, nine years old, used almost no spoken words in ordinary conversation. He still had not learned the word *cookie*, even after more than 10,000 trials.

In a radical change of approach, his therapists decided to seek his engagement in varied contexts to determine if he would spontaneously ask for the name of something interesting, which he had spotted. On a California beachside sidewalk, sure enough, he noticed the word "sidewalk" painted on the sidewalk. He asked, "What's that?" through gestures, and his therapist, delighted by the communication, said, "sidewalk." The boy's first attempt was a weak imitation, but the therapist encouraged him further by saying, "That's right . . . sidewalk." *Bingo!* On his second attempt, he accurately said the word *sidewalk*.

With a tricky new mix of exploration, encouragement to show interest and initiative, the escape from a training situation and environment with a long history of negative emotion, and newly positive emotional engagement from both child and therapist, two trials worked marvelously. Later in this book, we will encounter many similar instances for positive changes in individuals, organizations, and varied groups.

## A Fishing We Will Go

And now a fishing tale or two. When my daughter was two and a half years of age, we regularly visited a nearby lake to go fishing. My first goal was to introduce Leilani to the mysteries of the lake, forests, and all of the natural world. A simple strategy would have been to just sit on the shore and wait to see if fish would surface, if ducks would swim by, if red-tailed hawks would circle, and so on. But we chose the more complex approach of doing all that vigilant watching *combined with* a hook plus a worm plus a floating bobber fish-biting indicator. Now, we had many pieces of a truly positive, excellent tricky mix. We were in a position to secure a fabulous mix.

As we continue in any context and activity, there is the potential to build up an increasingly good mix through the accumulation of experiences and expectations. For my young daughter and fishing, the experiences and expectations began to build soon into our first fishing trip. The line was out in the water, the hook and worm sank out of sight underwater, and the indicator was bobbing on the surface. So what happened?

The beauty of the situation is that neither I nor Leilani knew what would happen next. We had no choice but to wait and see what emerged. When the bobber started darting around and then dipped under the water, we pulled the rod to hook whatever might be there—a small fish, a big fish, or perhaps something bigger, like the Loch Ness Monster, or whatever. Our first fish was a real fighter, despite its modest size—a flashy Bluegill that we managed to reel in together. We plopped it into our take-home tank, a spare summer picnic cooler. Then we watched it swim around with its sparkling colors and high vigor. Leilani's eyes were wide with wonder at our trophy, now ready to go live in our pond back at the farm. The whole event just delighted her.

Now the dynamic mix was transformed. As we continued on to the next cast of the bobber and line, she was no longer the same—her emotion had up-shifted to high positive and so had her expectations. In turn, her attentional and pattern processing was enhanced.

"Daddy, Daddy, let's get it!" she cried out whenever the bobber moved from lazy drifting to frantic movement. All these aspects of an ongoing highly positive and challenging mix continued as we caught six additional fish, stored each in the cooler, transported them home, and released the lucky travelers into the large expanse of our pond. In consequence, her later memory retrieval of event details was enhanced, and she was an even more skilled fisher when the next fishing trip came around.

Four months later and five additional trips to the lake to fill up a cooler with more Bluegill, and this two-and-a-half-year-old and I had accomplished good stocking of our farm pond. Far more importantly, Leilani and I experienced the kind of cycle of interaction that is a landmark of positive growth experiences for young children, older children, adults, and organizations alike. As the cycle went on, we each found more and more challenges, more skill acquisition, more positive anticipation of the next episodes of fishing and parenting/coaching, increased confidence, an expanded set of highly retrievable and relevant long-term memories, and more laughter and fun.

Coming up, you will encounter one particular fish that was caught in our pond a few years later. This largemouth bass splendidly lives up to his name. Imagine a frog in the water, sitting near this fish—something akin to the view of that big mouth in the photo would be what the frog would see just before he's about to be gobbled up. Or not! The frog may escape! What happens between predator and prey in these situations illustrates the complex *dynamic tricky mixes* that we will explore throughout the book, for unfolding events in nature and in all manners of events in human society. For the moment, with regards to this bass, know that after his brief photo session, he was released to swim freely and hunt again, back in his familiar watery territory.

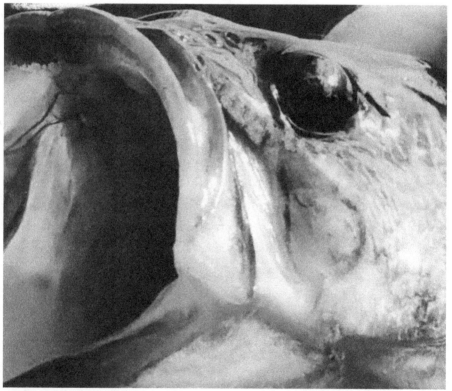

Just one of the numerous, lively, hungry fish we have enjoyed catching and releasing at our PA pond.

## Preschool Resources Wasted vs. Fruitful

Now we will take a look at preschool situations where resources were poured in with high expectations of success, yet where we can document persistent failure across a great expanse of time, as well as truly shocking and heart-warming advances. We will see a stark contrast between unsuccessful attempts at tricky mixes and dramatically positive dynamic tricky mix success.

## The Wasted Preschool Attempt

One year of preschool for three-year-old children was set up in Head Starts, with confidently predicted achievement of higher language goals. However, rigorous assessment showed absolutely no language benefits by the end of the year.

## The Positive Tricky Mix of Conditions

Immigrant children with no English skills were placed in preschool with kids who were native English speakers and where the tricky mix included active back-and-forth conversations between the immigrants and nonimmigrants. In just one year of this tricky mix environment, preschool immigrant children often leaped ahead, and their English skills matched those of the children from monolingual English-speaking parents.

In line with Dynamic Systems theory, the subset of children who showed these great leaps forward was prepared for the school years ahead and was the one that also encountered multiple positive convergent conditions. They were immigrant children who were socially outgoing and directly engaged with the English-speaking children, where positive emotions and fun were shown by both the learners and their non-immigrant conversational partners. It was noted that activities were frequently shared and in the early months, signs of progress led to snowballing effects in future months. These dynamically improving snowballs consisted of increasing English progress, increasing confidence and high expectations, and increasingly complex and challenging back-and-forth exchanges between the learners and the native English-speaking children.

# What New Dynamic Mix of Conditions Would Lead the Mighty Bald Eagle to Yield to Little Songbirds?

Below, you will see an unusual image of a bald eagle. On a flyfishing expedition along beautiful Pine Creek in central Pennsylvania, I was delighted to see a bald eagle landing high in a tree beside the trout-filled water. I raced to find my camera and began taking some pictures.

This bold, strong, confident eagle was quietly scanning the creek below for where the trout might be most abundant. Its white head glowed in the June sunshine, and I was able to snap a whole series of pictures. Just a few minutes before this specific picture was taken, what had been a comfortable situation for the eagle shifted as songbirds repeatedly approached the perched "ruler of the skies." Kingbirds and red-winged blackbirds dynamically re-mixed the conditions for this eagle. After studying this picture, reflect on what you think is going on.

If this eagle had only one songbird to deal with, it would have stayed on its ideal perch much longer. It could have easily protected its position with a swift move of its sharp beak or a sudden flap of one of its massive wings.

Instead, after a bit, more and more songbirds found the eagle and tried to intimidate it by diving dangerously close to it. Now, attacks were coming from all directions—from the right, the left, and directly overhead. As soon as the eagle turned its head and eyes to monitor the approach of one bird, another would dart in from a different direction. The eagle became frazzled, letting out a few threatening cries toward its attackers. And then

*Bald Eagle at Pine Creek in PA*

it just surrendered, and in this image, we see the eagle diving away as fast as it can while the little birds dive, too, right on its tail.

We will now extract a key metaphor from the scene. First, let's fully assess what the picture shows—a disgruntled bald eagle fleeing from the repeated dynamic "streams" of kingbirds, blackbirds, and other small birds, which are diving at the usually powerful eagle, the "emperor of the skies."

Metaphorically, similar patterns may play out in social-political situations of unequal power, instances where "small" ordinary citizens challenge power, dominance, and inequality. The "waters" or "attack streams" that ordinary citizens possess are inexpensive resources of truth-telling, persistent free speech against massive greed and abuse of power, mutual encouragement, votes, boycotts of products and places where corruption is rampant, and creative breakthroughs of organization and persuasion. These may all be dynamically used as powerful *tricky mixes* that force dominant, money-mad, shameless, polluting, exploiting "emperor" figures in any community or country to lose their power.

## Empowering Moroccan High School Students to be Changemakers for Education

High schools in Morocco have been stuck in an old paradigm of education where teachers are in charge and determine what will be studied and tested. The content and the skill goals are stuck in notions of what was expected fifty years ago rather than the changing landscape of careers and future opportunities.

A fellow under the international, well-established network called Ashoka pioneered a breakthrough approach. By directly fostering the roles and expectations of the high school students themselves, this one innovator is helping to create a dynamic systems-based snowball of successive changes.

Ashoka Fellow Adnane Addioui is changing the public school system in Morocco by empowering students to lead the change themselves. Through his "Tamkeen Initiative" training program, high school students develop the mindset of self-educators and active problem solvers as they launch their own solutions to the problems in schools and turn them into innovation spaces. Adnane seeks not to simply transform the students' mindsets on what experiments to try or understand their own capabilities. Instead, breakthrough changes are sought in the overall education system through working directly in schools and engaging teachers, school administrators, and parent associations.

"If you want to prepare people for the future, you need to have a future-proof system that is not focused on content, but on how people analyze and create new patterns and models . . . [At present] if young people want to do something, they are very much discredited because they are young and supposedly lack skills. Tamkeen means "empowerment" in Arabic. The idea of this program is to show how young people can provide solutions that others may not have thought of, starting from their own environment—their schools—and involving all their ecosystem—parents, teachers, school administrators . . ."

## Chapter 3

# NAVIGATION MAP FOR THE BANQUET OF CHAPTERS AHEAD

When you explore and expand your knowledge and awareness about dynamic patterns in any field of interest, you are also preparing to better understand other fields and the possible breakthroughs within them. This core notion runs all the way through this book.

Please compare my role as the author according to two, complementary metaphors.

First, as a tour guide, I have arranged a plethora of "islands of experience" for you to visit. You may visit these islands in any order, charting your own path. Or, it is also perfectly reasonable for you to enjoy the sequence of chapters from beginning to end, which I have laid out for you as one optional path. Other tour guides might select different places to visit. My claim is simply that—based on my own explorations of the many "islands" I am sharing with you—I feel it is highly probable that you will emerge from this tour with vivid memories, new insights and integrations, an awareness of new strategies for solving problems, and fresh confidence that you can be a significant contributor to diverse breakthroughs in the future.

My second metaphor is that I am your chef. Accordingly, I have prepared a lavish "banquet" of varied problems, people, issues, situations, events in nature, organizations in flux, and surprising episodes of all kinds. You will see, "taste," and absorb where things are truly stuck, where individuals or teams break through with new strategies at a breakneck pace of advancement, and countless other dynamic patterns. My hope and strong expectations are that the banquet ahead of you will stimulate your awareness, enhance your willingness to experiment and try new partnerships and fields of action, enhance your skills in observing and understanding dynamic patterns, and lead you to new optimism about what any of us can achieve.

In choosing so many different "islands" and so many different "menu items for the banquet," I have drawn from many sources. You will notice right away that I invite you to learn from comparing dynamics and "tricks" from scientific experiments, from unfolding patterns in nature, from business and entrepreneurship, from medicine and psychotherapy and rehab—along with narratives about my personal episodes and the escapades of children and adults around the world.

Every ambitious banquet menu has to be selective yet strive for richness. So, rather than trying to show you something akin to the "greatest breakthroughs of all time," my banquet of examples and episodes is meant instead to draw you in and surprise you, keeping you exploring, delighting, and sharing.

Another core theme across this book is the need to reduce simplistic thinking, simplistic messaging, and simplistic solutions. Whether we like it or not, *complex dynamic mixes* will be at work in our individual and group endeavors. Unless we bypass simplistic descriptions, analyses, and action plans, we will continue to see— despite the best intentions of many to the contrary—countless disastrous *negative mixes* emerge, often persisting unchanged across decades. Now, in 2021, we saw such disastrous patterns all around us in the form of local and global issues. At the global level, our *negative dynamic mixes* have yielded rampant over-population, destruction of biodiversity, global warming and weather chaos, the COVID-19 pandemic, startling and increasing inequities, brutal and counterproductive profit-gloating private prisons, massive toxic pollution of our waters, air, land, and bodies, along with global arms races and conflicts that draw outlandish expenditures on weapons and war rather than on constructive human enhancements.

We can and must do far better. Learning to create complexity-respecting, innovative *dynamic tricky mixes* at the local and small-scale level—changes you certainly could be central to—will be a powerful way to deal productively and creatively with the issues we confront at regional and global levels.

So, please, plunge in and find out where this banquet and this tour take your spirit, your imagination, your hopes, and your plans.

## Chapter 4

# BRIEF SKETCH OF DYNAMIC SYSTEMS AND DYNAMIC TRICKY MIX THEORY

The essence of dynamic systems approaches is that all systems—physical, psychological, learning, or innovation—involve multiple, complex, non-linear, fluidly- and rapidly-interacting components, factors, goals, feedbacks, variables, and parts.

In the case of human learning across the life span, here is a nutshell statement provided by Thelen and Smith:

> To understand the developmental process through which multiple parts continuously and fluidly influence one another, we must study the dynamic organization of cognition as a complex system and empirically discover its points of stability, instability, organization, and reorganization.

## Dynamic Tricky Mix Theory

We will see this theory applied many times, across different-appearing contexts and domains.

The "trickiness" of Dynamic Tricky Mixes Theory (DTM) is in getting all of the components of change lined up in real-time events. For events with positive outcomes, the goal often is an advance in learning or the achievement of an excellent or even stellar performance. As with Dynamic Systems overall, at an abstract level, the same kinds of processes are in play regardless of the "content" of the events under discussion. But of course, it is interesting to see how the particulars of complex events vary by domain. Thus, within an overall approach, we will look at how learning by infants, innovation in business, rehab procedures, and other domains must incorporate some domain-specific language—descriptors, types of data, variables, experimental results, contextual and individual differences, and the like.

We will see that the same DTM theoretical processes help make sense of what happens on a particular occasion and what happens when the same individual/ group is tracked over many related events concerning the same objective.

## Playing Offense in the World's Most Popular Sport

In soccer (football in some regions), ten players on a team try to advance the ball and score against the opposing team's goalie. Even for the best teams in the world, this is "tricky" because the ongoing stream of dynamic conditions is complex and fluid. An offensive team that achieves a series of great passes and dribbles, followed by a successful shot on goal, does so by creating a cascade of momentary dynamic positive mixes.

Consider for the moment a player in control of the ball halfway down the field. Their next move needs to take into account a wealth of information, including the position of each of their team's players, each of the defensive players, and their most recent movements. More than that, the player needs to consider the history of performance by all of the other players and retrieve—from long-term memory—the most likely movement/strategy plan for the other players to anticipate what may happen next. Finally, the player with the ball must focus on which basic moves to execute next and with what particular force and direction and speed.

Whew! You can see that the dynamics are highly complex. Some coaches and sports psychologists have estimated that truly skillful players will be taking in about one hundred pieces of relevant information per second! And this whole "dynamic stew" gets even richer because similar assessments and panning are simultaneously occurring for all the players on the defensive team. No wonder that for fairly long stretches of a match, there is a see-saw pattern of who possesses the ball before either team achieves any cascade of many momentary dynamic positive mixes, leading to the eventual scoring of a goal.

## World-Class Pianists in Uninhibited Performance

We can, of course, put on our headphones or earbuds and experience a first-class recording of a pianist in concert or in the recording studio.

Dynamically, however, our experience of the music may be strongly affected by actually seeing the pianist's every move on the keys and pedals along with all other physical movements and their facial expressions.

In Gothenburg, Sweden, in a small jazz club in the company of two of my best friends, I was drawn into the music by one of the wildest physical performances by a musician I have ever encountered. Lars drew amazingly compelling jazz from the piano, with accompaniment solely by a stand-up bass player. If you closed your eyes, it was thrilling, virtuoso jazz. And then . . . with eyes open, you were treated to a weird but synchronous gymnastics whirlwind. Lars flew off the bench, and then off his feet, to perform a handstand on the keys as he continued to play flawlessly. He gyrated back onto his feet, veering suddenly left, then right, then hopping straight up and down. Along the way, his face accompanied this all with an emotional, rapidly morphing range of expressions. Believe me, he could feel the music fully and, carried by his vibrant expressiveness, so could we.

Among other virtuoso musicians, the pianist Glenn Gould also accompanied stellar music playing with dynamic and idiosyncratic movements of his body. As with Lars, there is *tricky mixing* going on. When you thoroughly know how the dynamic pattern of notes should unfold, you find new and personal ways of ensuring that the emerging music has full emotional nuance by:

> *dynamically mixing in extra movements of your body*
> *and achieving the deep experience of a flow state.*

## Bach's Double Violin Concerto as Adapted for Foundation for the Thrilling Modern Dance "Allegro!"

One of the most thrilling "arts evenings" of my life was spent in a small theatre in New York City. I was close enough to the stage to vividly see every move and every facial expression of the dancers in the Paul Taylor Dance Company.

As in previous sections, we may consider a basic, persistent underlying musical structure provided by Bach. Yet in Dynamic Systems terms, the performance I heard and saw was extraordinary because the experience of hearing the "duet" of the two violins was transformed by not only the complex dancer moves on stage at breakneck speed, but also by the rhythms created by their feet and bodies pounding the underlying wood. Further, the particular musical structures unfolding in live performances were inter-articulated with all aspects of the choreography. Overall, all dancers and players performed at a virtuoso level and with explosive energy and feeling. It was a tornado—but with far more grace. It was a brilliant sunset—but with all the shifting colors generating synchronous cascades. Unquestionably, it was a *breakthrough!*

This performance of Allegro forever changed my mind about the expressive power of dance. Dynamically, it was breathtaking in its originality, fluidity, beautiful sensuality, and its power to emotionally and intellectually engage. It provides a stunning example of how a concrete—and historically well-established—musical score does not predetermine the music itself. Instead, it is only the starting pointing, a springboard for endless, ever-emerging creative interpretations, extensions, and inspired live performances.

## The Importance of New Pattern Analyses: Re-Looking, Re-Interpreting What You See, and Embracing Other Viewpoints and the Considerable Ambiguity

Breakthroughs achieved in any area or domain usually won't be sudden, singular events. Rather, in *Dynamic Systems* terms, there will be a series of explorations, new mixes, further re-mixes, and integrations before success.

The strategies listed in the heading above represent a sample of strategies that often contribute greatly to breakthrough processes and breakthrough events.

In the next few pages, you are invited to look at a short series of photographs from my own explorations of the world. Each photograph is an invitation to go beyond your first look and first interpretation. Generate as many hypotheses as you can about what is "really depicted." Also, actively explore pattern resemblances—what other patterns in your own experiences are similar to those in the photos.

Please delay reading my brief text comments after the last photo until you have had some fun dealing with lots of possibilities and connections for these patterns. For a while, experience being a *dynamic tricky re-mixer*! Absorb new visual patterns, actively compare them to others, and look for extensions and variations, dynamically connecting them to patterns you already know.

*Just a simple Greek sunset?*

*Just a simple resting heron?*

*Just a fox at the simple, obvious entrance to his den?*

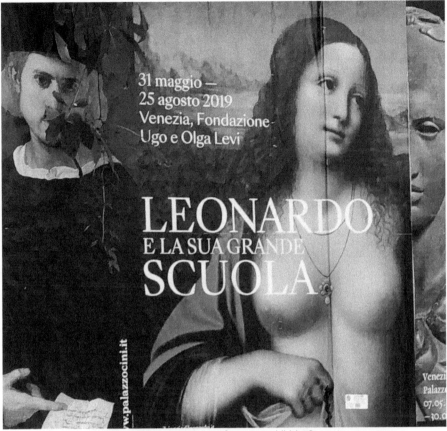

*Just a simple poster for an art exhibition?*

## Author: A Few Dynamic-Tricky-Mix Reflections on the Pictures

For now—just sit a little while with your own reactions, interpretations, and re-interpretations of these pictures. Continue to have fun exploring them.

Then scroll down to my comments.

Simple sunset scene? Nope. Always look beyond your first glance. The view here is looking west from the island of Santorini, across a bay of water. Beneath that water is a "caldera" bowl shape, created by an intense volcanic eruption in 1646 B.C. The resulting dynamic mix included an explosion that was estimated to be one hundred times more intense than that of the famous Pompeii eruption. It produced so much ash that the local town, Akrotiri, was buried. Further, a massive tsunami wave emerged which traveled seventy-two miles to the island of Crete and devasted the Minoan culture there—not only by killing thousands but also by destroying their trading fleet and towns.

Simple resting heron? No, this beautiful green heron is far from being in a resting state. More likely, he is in what we label a bit later in this book a "flowing state." All of his senses are on alert, to detect fish in the pond below, which will trigger a dive—deadly for the fish. Beyond that, we know that these green herons are tricky mix experimenters! So, if they don't see any fish, they will try to lure the fish out of hiding—to check out "lures" the bird drops on the water's surface in the form of small pieces of leaves or twigs.

Simple entrance to a fox den? Well, remember that foxes have a reputation as extremely elusive, fabulously tricky. What if several generations of foxes kept mixing and re-mixing the design of a large den, each year adding new entrances and exits? That would explain the discovery of some fox dens with a large underground central chamber containing fifty to one hundred different entrances. For such a den, any fox being chased by a human or non-human predator could be tricky indeed—perhaps mysteriously disappearing into some alternate entrance any time they find the path to a closely watched entrance blocked!

How about that simple claw rising up from the water? It turns out that a much more complex event was underway, an event we will later unpack and reveal. In a nutshell, though, the claw belongs to one of the two snapping turtles gently rolling around in the water in a sort of "mating ballet."

Thoughts beyond this being a poster simply announcing a Leonardo exhibition? Da Vinci is a suitable "poster boy" for lives that are lived as active dynamic tricky mix experimenters. He constantly explored and placed reminders in his monitoring notebooks concerning his adventures—new contexts, domains, and questions. Then he dynamically compared and re-mixed patterns from all of those varied domains. On the basis of this approach to life, he made tremendous lasting achievements in painting, sculpture, technology, science, medicine, and more.

# Chapter 5

# TWELVE DYNAMIC PUZZLES

An emerging theme already in this book is that you, your kids, your friends . . . everyone can become a productive "tricky mixer." With your attention and the coaching you're about to receive across varied situations, you will be able to spot what the complex dynamics likely are for any new situation. And, then, you will be able to generate likely new tricky mixes that help someone who is *stuck* move forward in surprising and encouraging ways.

In this chapter, you will have an opportunity to come to grips with some real-life narratives of folks who are *stuck* and for a while, feel pretty hopeless. In each case, try to generate some new moves/strategies that might move the person from stuck to achieving excellent progress. Likewise, for other puzzles I present, do some active thinking on how solutions to each puzzle can be found.

## Case 1. A Professor with a Myriad of Ailments

This professor faced what most professors do: considerable pressure as he strived for enough research and enough publication and enough recognition from peers to ensure his achievement of a tenured position. By his own accounting, these pressures dynamically mixed with the neglect of sensible, healthy meals. These components further led to antacid prescriptions in an attempt to control irritable bowel syndrome (IBS).

Well, our bodies and minds are very complex systems. So you may not be surprised that things then went from bad to worse. He developed persistent sinus systems, took a round of antibiotics that seemed to help, stopped the antibiotics, and the symptoms recurred. Then there were more antibiotics. This repetitive cycle continued. Even worse, additional infections and periods of low energy became frequent, and gut and bowel problems recurred. All of these negative aspects were cascading and colliding with each other, persisting for thirty years.

OK, reader. Now, it's your turn. Make predictions about what re-mixes of conditions would be sufficient to put this professor on a path to overall good health. Or do you think he has simply been dealt this fate—the kind of body that cannot fight off ailments, that he will always be *stuck*?

## Case 2. A Poet in the Catalan Region of Spain

This case concerns a medical doctor and the puzzle that arose regarding his father. His father was enjoying his career as a poet and professor in Spain. But one day, he suffered a stroke, and suddenly, the poet could no longer talk, write, walk, or even crawl. The rehabilitation his father received seemed to have no discernible effects on these life-limiting symptoms.

Because this poet was *stuck* after the stroke, despite many months of extensive rehab, it would easy to draw the simple conclusion that his brain was beyond recovery . . . unless there is a new mix of conditions, recovery seemed extremely unlikely. So, again, it is your turn, reader. What would you try in terms of a radically different, new remix of conditions?

## Case 3. Refugee Fleeing on Foot from Nazi Germany in Jeopardy

Imagine you are carrying scientific nuclear secrets as you and your wife frantically flee Paris, trying to reach ship transportation out of Europe. You must take rural roads and cross-country paths to evade the Nazis. Worse, your earlier soccer (football) injuries are acting up, and at times, you cannot move without severe pain. Your progress is slow for these reasons but also because you must give your wife considerable assistance because, from early childhood, she has been unable to walk or run at normal paces.

Eventually, this scientist escapes, survives, and flourishes, moving beyond wartime and experiencing significant career changes that contribute to his own and others' abilities to become *unstuck*, free from chronic pain. The man's training was in physics.

Your turn again, reader. Speculate on whether the pain treatment *break-throughs* this man pioneers were based in physics research or in entirely new arenas and perspectives.

## Cases 4 and 5. The King of England and a Male Superstar of India's Bollywood Movies

Each of these gentlemen needed a *breakthrough* that would allow them to over-come a malady that had persisted from about age four, right into adulthood, affecting their crucial roles.

So, reader. What was the shared malady, and what were the dynamic mix routes to becoming basically un*stuck* from the malady?

Here are a few more clues. Both of these guys were teased as children because of their malady. They often avoided social situations because they were too anx-ious about their likely performances.

You are correct if your answer was either "stuttering" or "stammering." Two words that refer to the dysfluency that arises in speaking, including, for example, repeating the first syllable of a word multiple times as in "par—par—par—parking."

For the moment consider two aspects of their situation: first, both men, as adults, had learned methods for limiting how often they stuttered. But they had learned different paths to the new dynamic mixes that worked. Again, see if you can anticipate what we will reveal later in the book about their approaches.

Second, not only were these two stutterers, but most chronic stutterers share some peculiar characteristics. A common assumption had been a too-simplistic one—that speech breaks down right at the moment of forming the sounds, that the speaker gets *stuck* on a sound at the level of motoric performance. However, there are multiple dynamic tricks that demonstrate something akin to the situa-tion with our dogs, Sail and Banjo. The dogs could no longer climb simple steps, but it wasn't because their motor systems were broken. Change the dynamic mix of conditions, and they move just fine. Similarly, when the dynamic mixes are changed for a stutterer, their motor production of speech sounds is flawless—as when they read a written passage out loud along with other speakers who are fluent or when they read a passage with sound-blocking headphones that prevent them from hearing their own spoken outputs!

Keep thinking about these two guys, the famous Indian actor Hrithik Roshan and King George of England. Later, we will identify the dynamic mixes that worked for each of them and relate those positive outcomes to other remarkable successes for people in all kinds of contexts and careers.

## Case 6. Two Drawings: Each Created in Five Minutes by a Preschool Child

Inspect these and reflect on these. Then make a prediction about what mixes of circumstances would lead one child to make such a simple drawing and another child to make a far more complex, realistic, and expressive drawing. Later in the book, we will explore this basic question, not only for today's young children, but for children and adults form prehistory times and onward.

WHY IS CHILD ON LEFT STUCK ? ?

2 DRAWINGS OF PERSON

List all possible explanations you can present for why the 4-year-old on the left is stuck at a simple drawing level, and why the other 4-year-year old draws with so much more complexity and expression.

*Two Contrasting Drawings of a Person.*

Related to this puzzle is the question of genetics. How likely is it that innate, biological gifts would underlie one child, who races along with drawing/art complexity. If innate advantages are bestowed on a child, what, more specifically, are the child's characteristics that feed into superior performance in art? Be as specific as possible.

Similarly to drawings of people, most three- to five-year-olds also are *stuck* when it comes to drawing houses or any other objects of simplicity. The children already know what real houses look like. So why do so many get *stuck* at very simple levels of drawing? Again, as above with regards to drawings of people, make a prediction about what mixes of circumstances would lead one child to make a simple drawing and another child to make a far more complex, realistic, and expressive drawing.

## Case 7. Lightning Strikes and Tennis Strokes

The dynamic puzzle we face here is what happens after this tennis-playing engineer is zapped by lightning, just as he is setting up his tennis serve during a match. In his own words:

> As I raised my arm to serve—crack, bam!—a bolt hit me on the head. As I fell, I heard a sharp snap from my tennis shoe on the asphalt surface. Sprawled out on the court with rain pelting my head, I struggled to regain consciousness from what appeared to be a very bad dream. I glanced across the court to see my partner scrambling to his feet. This scene was not a dream. We had been hit by lightning!
>
> We picked up our gear and ran for the shelter of a porch at a nearby recreational building. The rain and lightning from the thunderstorm intensified as we set a new world's record for getting off the court to safety. Tony and I checked each other out. As best we could tell, our bodies appeared to have no physical damage. We speculated that the electrical pulse must have traveled down the path of least resistance, our body sweat. After a few minutes, Tony decided to leave. I realized I was not fine, feeling dizzy and brain-fogged.

So, readers, what do you predict? Over the next few years—with this guy, Jack—will he be *stuck* with brain fog and other limitations? That is, *stuck*, sort of like my dog Sail at one sad point, where even a few stairs were too much to climb? Or will Jack find some new dynamic mixes—and if so of what sort— that lift him toward a renewed mind and spirit, which will feed successful life trajectories?

## Case 8. The Sight of a Blind Guy

A guy with ordinary vision is driving an ordinary car down an ordinary street. Explain how his sighting of a blind guy with a cane, who was walking along on a nearby sidewalk, could dynamically cause a major life *breakthrough*.

## Case 9. "Yestermorning, I Maked a Tiger."

A two-year-old girl produced this little gem of a sentence. Discuss why this imme- diately makes sense to adult English speakers. Further, explain why this example gives crucial insights into how today's children learn, as well insights into how creative *breakthroughs* in symbol use by our ancestors more than 300,000 years ago were crucial to the evolution of increasingly complex brains.

Related puzzle: Why did a three-year-old boy say, with great conviction, "I can't eat the baby peas!" What components of his prior experiences were dynami- cally mixed to create a very non-adult interpretation of a forthcoming meal?

## Case 10. Simplistic Proposals Concerning Human Language

Some scholars in ancient Greece and some modern linguistic professors have claimed that all children, regardless of culture and the dominant language spoken therein, display an extensive set of language structures by six years of age. They have further claimed that the explanation behind this observation is that there is a human-specific, innate basis for language. Noam Chomsky made a metaphoric comparison to human organs, such as the lungs for breathing and our hearts for blood circulation, and proposed in 1964 that the only possible explanation for human success with language is an innate "organ" for language.

Speculate on how Dynamic Tricky Mix Theory might provide an alternative framework, which not only deals with the typical successes of young children regarding spoken language, but also provides explanations for a broader and more complex set of observations: sign language acquisition, individual differences, and why and when some children are delayed—sometimes seriously—with their language acquisition.

## Case 11. Early Hand-Held Tools—Called Hand Axes— Piling Up On an African Lakeside 600,000 to 900,000 Years Ago

In the short life spans typical for prehistoric Africa, why would huge amounts of time be invested in making thousands of hand axes from stone? And why were these axes shared in just a few locations around a particular African lake?

In India, as well, there are mysteries to be solved around hand axes from a period spanning from 600,000 years ago to 1.7 million years ago. Early Paleolithic quartzite "hand axes" served as tools for cutting bones and as hide scrapers and small hammers. Specifically, people want to know the dynamics explaining why so many hand axes were available for a fairly small number of

*Hominins*—primates closely related to humans—near the town of Attirampakkam in India.

Further, there is a gap in the record—we lack relevant remains, so we don't yet know which *Hominin* species were the inhabitants, the toolmakers, or the tool users.

The hand axe you see in the next figure is from Isapur, in India. Reflect on what symbolic meaning this stone object may have had for ancient clans, aside from any significance as a possible tool.

*Prehistory Handaxe from India.*

## Case 12. A Lopsided Superbowl Score

Imagine a very special "Superbowl above all Superbowls" takes place, and the final score is amazing. The winning team scores 210 points, and the losing team, a measly seven points. Identify who the likely winning and losing teams are in this mythical matchup. Also, explain as best you can why the score is so lopsided.

## Chapter 6

# WHY THE F\*\*K NOT?

Oops, did I lapse into profanity? Yes but with purpose.

Here's a remarkable case where the occurrence of important changes for a ten-year-old rested squarely on a *dynamic re-mix* of conditions. A simple new obsession with words, such as the "F word," and rap rhythms in popular songs led to a positive cascade of new achievements. This mother tells the story in rich detail:

> My ten-year-old son starts to rap after someone on YouTube says that Eminem is autistic.
>
> "I don't know if Eminem is autistic or not," I tell my son. "YouTube says lots of things."
>
> "He looks worried all the time." He tilts his head, holding the worn-out headphones in place. "He writes exactly what I think," my son shares.
>
> "He writes about stapling his teacher's testicles to the desk," I say.
>
> "I love him," my son says. "But the radio edits make me nervous. I wait for the empty spaces, and this bad feeling builds inside me, knowing they're coming and just waiting for them."

Suspense in any form is unbearable. He shivers and opens and shuts his hands, fists to flat palms and back. Over and over until I place my hands on his.

"Don't tell your dad."

I download the explicit albums. They contain graphic sexual acts and lyrics about abuse and drugs, which he has never heard of. To my relief, he barely notices the actual content. He just loves the profanity. Every "f**k" makes him chuckle.

"If you teach your little sisters any of this, I'm deleting the albums," I warn him.

"Hee hee!" he cackles, as Tommy Lee whoops Pamela's a$$. "I used to think it just meant donkey!"

## Pause Button for Readers

Please take a few minutes and write down reasons for/against this mother allowing her ten-year-old son with autism to pursue this raunchy music.

## Resume Button for Readers: What We Know of What Happened Next

Is it still misogyny if the listener is unaware of the context? I do not bring up girlfriends locked in trunks or wives getting murdered. I step back and listen to the music with a ten-year-old's ears. Magical four-letter wordplay.

My son hears himself in these lyrics. The isolation, anger, and frustration about authority figures and expectations, about who and what is acceptable, about what it means to deserve love or even basic acceptance. The messages are wrapped in clever rhyming schemes that he can touch and manipulate. It distances itself from his overloaded heart with its playfulness, allowing him to think and feel without becoming frozen by panic. It allows him to talk about his experiences, even day-to-day interactions.

"Read me one of your poems," I say.

My son's raps follow the exact same rhythmic patterns as Eminem's, but he creates his own themes and poorly spelled word combinations:

*Geomitry's answers are defe-dently not forbidden*
*and you can trust me when I say that there truth is not hidden.*

The confines of an established, successful form give him the confidence to explore his own world. He shared his new words at school, and the teachers loved this new hobby—at least at first. His classroom assistant asked him to sign her cellphone case.

He begins to think, learn, and demonstrate knowledge in rhyme more often than not. He bursts out in math with long verses about angles and the formulas for area and perimeter, and before long, his classroom teacher is torn between admiration and irritation. He is smarter than her other students. He is also louder.

"That was very good," she says. "But can we share at the end of the day?"

I get daily emails:

*Your son cannot stop rapping.*

*I need your son to stop rapping.*

*I admire your son's commitment to rapping, but it is very difficult to teach the parts of a plant cell while someone is loudly rapping about mitochondria in the back of the classroom.*

"Poetry is my special interest right now, Mom. You're a writer. You should understand."

"Yes, but I don't yell out passages from my novel while a teacher is explaining the differences between a cell wall and a cell membrane."

"I drop subject-appropriate rhymes. I'm listening and learning, and she's mad?"

His point is a valid one.

"There are rules we follow in school." I cringe as I say this. I am also a teacher, but it's different when it's your own kid.

## Pause Button for Readers

What would you suggest as ways of resolving the conflicts in perspectives and strategies faced by this parent and the boy's teachers?

## Resume Button for Readers

A grunt. He walks over and sits down in front of his laptop. He opens it, cracks his knuckles like he saw Snoopy do in a Charlie Brown cartoon years ago, and starts tapping at the keys. I tap on my own laptop across from him, sneaking peeks as he alternates between grimaces and smiles. Listening to him mutter, pounding on his backspace button, revealing an imagination I thought he didn't have. An imagination he never disclosed until he started writing.

After forty-five minutes, I have written half of a chapter that I will probably throw away. It's garbage. My son prints a page and hands it to me, whispering along over my shoulder as I read expertly rhyming lines about small-minded teachers and mothers who stand up for free speech. He writes jokes within the lines, utilizing the sarcasm he struggles to understand in real-life discussions, expressing himself through words that he cannot say out loud unless he has written them down first.

"You should write my novel for me," I complain.

"No, Mom. That's your special interest. I can't wait to see your book on the library shelf. I will bring all my friends over and point to it and say my mom wrote that."

I do not cry.

"Then I'll write a rap about you, just like Eminem does about his mom!"

Wait a minute.

A wild laugh, his head thrown back, he takes off up the stairs. Rhyming love-insults fly over his shoulders toward me. His laptop is held tightly to his chest.

He will not be able to rap his way into college or even into receiving an invitation to sit at the lunch table with the other boys in his class. And I wish he'd picked almost any other artist—at least for now—to emulate and idolize. But this is my son communicating, sharing what he knows and feels with the world around him instead of holding it inside, and it's Eminem's music that was the catalyst.

Will we need to have a larger conversation about these songs and their meaning at some point? Do I worry about him internalizing messages that a ten-year-old doesn't have the tools to deconstruct?

Yes.

But I also know that my son has learned that there is power in telling his story. He sees others listening to and understanding his words, and his desire to speak and share—to interact with the world—grows in turn. The last six months of his rapping have changed his mindset, which is no small thing for an autistic child. Together, my son and Eminem built a bridge. I will enjoy every moment of him walking across it, my ears burning, my son's heart on brilliant, rhyming display.

This mother, Hannah Grieco, is an education and disability advocate and writer in Washington, D.C. Her work has appeared in *The Washington Post's* "On Parenting," Motherly, *Arlington* Magazine, *The Lunch Ticket*, the Hobart Pulp blog, and more. Her account is used with permission.

## Brief Summary of This Case in the Framework of Dynamic Tricky Mixes

The dynamic cascade of changes, which positively worked together, is striking for this boy. His enjoyment of Eminem was accepted and folded into new interactions with his mother, peers, and teachers. These interactions gave him more acceptance, more recognition of his intelligence and creativity, and more incentive to keep interacting richly with others. As part of this *dynamic re-mix of multiple conditions*, as his mother notes, is an entirely new mindset and set of expectations, which influence nearly every learning situation and social exchange this child with autism enters.

Chapter 7

# SHOOTING SACRED COWS &
# RE-MIXES THAT ASTONISH

I n the examples so far, it is clear that moving beyond a lousy mix of conditions to a more positive tricky mix often involves identifying and eliminating long-held assumptions and/or routine procedures.

First, we will illustrate some sacred cows that have held children back from rich opportunities for development. Following those, we will examine ways that we can increase our success in identifying and "shooting," "turning out to pasture," or otherwise removing whatever sacred cows we find in human endeavors.

## An Infant Who May or May Not Become Bilingual

On a gorgeous summer day in June while in Sweden, I had a revealing conversation with a couple about their two-month-old delightful daughter. It seemed that this loving couple—an American man and a Swedish woman—were in an ideal situation where each was fluent in the other's native language. Distressing them, however, was a sacred cow that was so sacred they didn't even know it was a cow.

Many experts on language had told them that it was essential that their baby hear Swedish only from the mom and English only from the dad—only the native language from each person Otherwise, the baby would flounder in language confusion and face terrible long-term developmental problems. This kid would not even be in the one-word stage for many months to come, yet they were frustrated and on edge.

"The key problem," the mom said, "is that sometimes we want to communicate in just one language to each other but while the baby can hear. If we did that, it would mean Sophia would hear what she should never hear: me speaking English or my husband speaking Swedish! What can we do?"

I wanted to help. And I knew I could. That's because the "sacred" advice they had been given was not based upon credible evidence but upon mere assumptions. Moreover, consistent, convergent research results were already available that showed that children in the first three years of life do great with acquiring two languages and becoming fluent bilinguals if the mix of conditions includes highly fluent speakers who bring high responsiveness to the child and high positive emotion into the family conversations. So, on the spot, I *shot* their sacred cow and turned them loose to have open, flowing conversations with each other and with the child. English whenever they felt like it. Swedish whenever they felt like it. Sophia would sort it out just fine and acquire two languages at the same pace that a less-lucky child would acquire the one and only language they hear.

## Sacred Cows Concerning What "Science" Can Address

If we go back in time just three hundred years to the year 1700, it is easy to identify how widespread sacred cows were. Whole "pastures" of sacred cows were based upon European religious beliefs and biases. Accordingly, it was considered either uninteresting or heretical and sinful to search for scientific accounts of any of the following three topics:

1. The long history of changes in animal species, plant species, and planetary physical features
2. Whether profoundly deaf individuals could acquire complex communication, thinking skills, and knowledge
3. Effects of human choices on economic prosperity

These areas were off-limits for scientific investigation because religious frameworks had already judged the issues and established what were considered correct doctrines. And these prohibitions on inquiry held sway even though other truly incredible advances in scientific methods, topics, and conclusions had been achieved between the years 1400 and 1700.

So, by the year 2000, were sacred cows no
longer such a danger in modern
societies?
Far from it!

Another modern example is the contrast between studies of economic processes and outcomes versus studies of human happiness.

Scientifically trained economists are rampant in universities, government, corporations, and well-established institutes. Fierce debates between different perspectives rage. However, all believe that financial data should be systematically collected and analyzed.

Scientists studying happiness, on the other hand, are about as rare as hen's teeth. The two central sacred cows about happiness are that it is too complex and too personalized to be understood in any scientific fashion. But wait a minute—those same limiting beliefs used to hold concerning economic phenomena or human personality.

The Happiness Institute in Denmark and two university institutes at Berkeley and Illinois provide a forum and focus for "heretical" investigators who believe that happiness data should be systematically collected, analyzed, and interpreted. Moreover, the framework provided is more open than most fields of scientific inquiry compared to Eastern and indigenous cultures and Western mainstream cultures.

Happiness, of course, does matter to all of us. The Happiness Institute provides revealing observations about how certain cultures succeed in creating *tricky mixes which support high happiness levels*. See more about this under the section "Flat, Flat Copenhagen Installs Their First-Ever Ski Slope: A Lesson in Freeing Dynamic Idea Flows in a Corporation."

## Educational Equal Opportunity & Achievement

It is commonly assumed that in the United States there is a "flat" world when it comes to education. That is, everyone knows about all the options, approaches, finances, and then children regardless of neighborhood, income, ethnicity, parents' education, and so on receive similar and effective education. It is assumed that there are only minor variations from this scenario.

Ha! The realities of educational disparity are stark in the United States as well in most countries.

The actual mix of conditions varies dramatically by community. This holds true even though there is a strong similarity in the available components of schooling that might dynamically mix together to create engaging, challenging, highly effective learning and educational persistence for children and young adults.

In New York City, it has long been possible to find schools in impoverished neighborhoods that achieve negative, toxic mixes of educational conditions. Jonathon Sokol and others have documented similar depressive charades of education in other cities. In New York, one research program looked at schools where the graduation rate and school attendance rate were shockingly low. Given these numbers, it is clear that the schools were failing most of the children. However, some worried that the children's parental support was so poor that the children could not succeed in any school.

This false conclusion was negated when a dramatic re-mix of educational conditions for the same children was put in place. The results were truly astonishing. Now the graduation rate and school attendance rate are excellent. The children, despite being *stuck*, had been ready all along in terms of prior knowledge, cognitive capacities, and motivations for learning and success. All that they needed was a radical re-mixing of setting; teacher attitudes, expectations, and strategies; and pathways to mastering academic material.

So, what was the dynamic re-mix in this instance? It was forming a charter school utilizing boats in the harbor along the Hudson River in New York. This Harbor School integrated math, writing, science, reading, literature, and other domains around marine-related topics.

Students often worked in teams and were strongly supported emotionally with flexible coaching strategies and an emphasis on the fun and challenges of learning by their carefully selected teachers. The students discovered their own high capacities for learning and dedicated study, and as time went on, they saw more and more clear demonstrations of just how capable they were. In consequence, their positive expectations of learning kept going up and up—dynamically feeding into even stronger paces of learning and an eagerness to attend school, work hard, and engage fully with their teachers and fellow students.

## The Transition to College-Level Education

High schools provide further evidence that the realities of educational disparity in the United States are stark.

When students are poorly prepared by their small and rural high schools in terms of mathematics and science, the students often crash in their freshman and sophomore college courses. Joint research at the University of Texas and the University of California, Berkeley began by documenting the negative performance in freshman calculus. The average grade was a D! The majority of these students were failing, and this was particularly true for freshmen from small and rural high schools.

So, what could be an effective dynamic re-mix in this instance? Without changing the students' past courses or their basic cognitive abilities, would it be possible to radically re-mix learning conditions so that these students would master calculus?

Step one toward re-mixing was to shoot another sacred cow—*that professors should lecture most of each class*. Absolutely, this is the standard approach for calculus. But the University of Texas and the University of California re-mixed this so that out of a fifty-minute class, only ten minutes was spent to lecture on new material. For the remaining forty minutes, the professor roamed the class and coached pairs or small groups of students, encouraged and laughed with the students, answered questions, and so on. Now the class time became primarily interactive. The professor and the students were socially and emotionally engaged, responsive to each other, and comfortable pursuing a topic or problem until it was well clarified. Odd! Unusual. But satisfying to all. And as more and more of these re-mixed class sessions occurred, positive expectations and willingness to contribute exploded and fed into the dynamic mixes of the next class sessions.

There is also more to the me-mixes for these college students. Students and their professors made time for shared social occasions—softball games and other events purely for fun and bonding. Social connections strengthened. Further social bonding occurred when students were required to always do their homework in pairs, with each pair member encouraging and helping the other until both agreed that they had adequate solutions to the homework problems. In combination, all these re-mixed conditions contributed to positive attitudes, expectations, and emotions as well as low anxiety both during classroom learning and homework learning opportunities.

As in the cases of the children in prior examples, these college freshmen had been ready all along in terms of prior knowledge, cognitive capacities, and motivations for learning and success. They now passed calculus with an average grade of B, and most could easily go on to enter more advanced math and science classes. All that they needed for college math success was a dynamic re-mixing of setting; teacher attitudes, expectations, and strategies; and pathways to mastering academic material.

## Flat, Flat Copenhagen Installs Their First-Ever Ski Slope: A Lesson in Freeing Dynamic Idea Flows in a Corporation

The story of how this remarkable ski slope came to pass is part of the bigger story of how a corporation can use a breakthrough design of its own processes to cultivate better urban building designs in cultures around the globe.

The Danish corporation involved here is named BIG. Its founder and leader is Bjork Ingels, and thus Bjork Ingels Group makes the acronym BIG.

By integrating local cultural values and aiming for high levels of visual excitement and functional effectiveness, BIG has created buildings with award-winning impact in China, Sweden, and China.

BIG has an advantage in that any new project has positive expectations from the whole staff because they have all been warmly included in the early phases of design for previous projects that went on to have high success. Ingels has brilliantly incorporated into the design process many of the dynamic tricky mix aspects we have repeatedly stressed as feeding into breakthroughs and avoiding dealbreakers.

First, ideas in early the phases are invited and recorded from all—not just from the top executives and designers—and shared with all for discussion and evaluation. Second, wild and creative ideas from non-expert designers often are *actually adopted* so that everyone knows they are included. Third, community and culture are understood at a rich level and thoroughly integrated to create original tricky mixes of design features for each project. Fourth, there is no rush through the early design stages—plenty of time is taken to re-mix, re-mix again, and keep the process open and evolving until a compelling plan emerges.

What emerged was Copenhagen's first ski slope—on top of a huge power plant—which opened in 2017. The overall tricky mix was a fully modernized power plant with a specially shaped and planned roof area. It incorporated parks, walkways, and the installation of a snowless—but still thrilling—set of downhill ski slopes. Further, the height of this new recreation area created fabulous views over Copenhagen day and night. One view of this architectural breakthrough follows.

*Flat Copenhagen Gets Its First Ski Slopes—on Top of a Giant Power Plant.*

The political decisions to support such a wild architectural breakthrough also fit smoothly with the ways in which Copenhagen, and Denmark overall, approaches community-sensitivity planning and governance.

As repeatedly shown through research, happiness levels within Denmark are reliably near the top compared to other countries in the world. Local, regional, and state decisions feed into widespread happiness. In effect, dramatically positive dynamic tricky mixes are maintained which support citizens of all ages and backgrounds in their autonomy, safety, common purpose, health, jobs, leisure, access to the outdoors, education, communication, easy and efficient transport, and mutual problem solving. This tricky mix prioritizes relationships and mutuality and economic equality over acquisition of great wealth.

*Copenhagen Community Provides Easy Access to Cafés, Beautiful Walking Areas, Boats at Harbor, and More.*

## Transformations of College-Plus Educational Experiences

As teachers and change agents for university students, Christopher Uhl and Dana Stuchl have made observations and recommendations that are relevant to all levels of education for children and adults. Here is one key summary:

"At present, education programs and policies give little heed to helping teachers, much less students, grow toward self-understanding, self-acceptance, and self-actualization. Contemporary schooling, more often than not, reinforces the "trance of unworthiness," positing that a student's worthiness is conditional upon her classroom performance. So it is that young people become, in a sense, like little Sisyphuses—always laboring uphill to get the praise, acceptance, and love they desire, only at some moment to slip back down the love slope, feeling unworthy and unloved."

But what if teachers created the space for young people to explore, to know themselves, and, in so doing, to discover their unique personhood? Specifically, what if four key messages provided to young people by their teachers were those on this list.

> You are real and you have a right to be here.
> Your questions matter and deserve
> consideration.
> What you are feeling today and what you
> have to say are important.
> You are fundamentally good and loveable
> just as you are.

This perspective is one I have used in my own teaching, exploration, and coaching activities. Uhl, Stuchul, and many others have seen how often students achieve self-changes that feed into progress in all their endeavors, not just learning while in educational institutions. From the dynamic tricky mix framework as well, it is inevitable that core changes in a person's characteristics will activate a cascade of new positive differences in thinking, caring, creativity, interpersonal connection, cooperation, and achievement. Here is a key conclusion from these authors:

"Self-awareness, personal growth, and the capacity for love and service on behalf of the common good become the ultimate goals. This is the change that teaching as if life matters promises."

## Their Credit Scores are Zero—Hey, Let's Lend Them Some Money!

Imagine being an adult in a poverty-stricken rural environment in India, Africa, or Indonesia. So far, just like your grandparents and their grandparents, by objective measures you are stuck in terms of your low expectations of progress and your minimal actual progress towards a more prosperous life.

Now, imagine you are a wealthy banker looking for new opportunities to achieve a good return on new investments. For most, looking to the world's poorest would seem highly unlikely to say the least. Perhaps to even approach that possibility a banker would first need to slay a sacred cow of "seek maximum investment returns."

Matt Flannery, founder and CEO of Kiva and Ashoka Fellow, emphasizes that Kiva connects people through lending to alleviate poverty by leveraging the Internet and a worldwide network of microfinance institutions:

> "Most of the world's poor lack access to sustainable financial services, whether it is savings, credit, or insurance. Microfinance was born from the belief that low-income individuals are capable of lifting themselves out of poverty if they are given access to financial services—all they need is the opportunity, and Kiva loans help provide this.
>
> This belief embodies the ideals that Kiva is founded on: that people are generous by nature and will help others if given the opportunity to do so in a transparent, accountable way; that the poor are highly motivated and can be very successful when given an opportunity; and that, by connecting people, we can create relationships beyond financial transactions and build a global community where we express support and encouragement for one another."

Chapter 8

# CLASSIC OVER-SIMPLIFICATION ERRORS

I magine a situation where you, your corporation, your foundation, or your school are keenly motivated to launch a new initiative toward a crucial goal based on an original and highly promising idea.

In most cases, available budgets and human resources will have clear-cut limits. So a simple and seemingly obvious route to follow is to research and evaluate many ideas and then choose a winner, the "best" idea from a pool of candidates.

Moreover—and, again, because available budgets and human resources will have obvious limits—in implementing the chosen idea, a popular and simplistic strategy is to devote all resources to carrying out new projects while monitoring how well the project is achieving goals. After all, you or your team have chosen the "best" idea. Only the best will do, and your success may feel inevitable.

## "Choose the Best and Leave the Rest" Strategy

What we have just described fits with this "choose the best and leave the rest" strategy label. It is an extremely popular strategy, but it carries with it many extreme shortcomings. Because there are complex dynamics that cannot be avoided at every stage in choosing, implementing, and monitoring an idea, "choose the best and leave the rest" often leads to highly disappointing outcomes.

In this chapter, we will review many aspects of the decision process for ambitious projects. And here we bring in another contrasting approach to the strategy just mentioned.

## Causal Experimental Evolution: Strategies for Re-mixing and then More Re-mixing

By grounding efforts to achieve breakthroughs in awareness of complex dynamics, this strategy points toward dismissing any "silver bullet" approach toward new achievements. Instead, in moving into any new project, there is an emphasis on setting up comparisons between multiple seemingly promising new ideas. In addition, care is taken to deploy enough resources to give valuable feedback about early-phase progress and dynamic processes and then adjust again and again until effective dynamic mixes are established.

Consider below a few well-documented examples of situations where the "choose the best and leave the rest" strategy led to many months or years of zero or minimal progress toward goals.

## Mathematics in Grades One Through Five

Testing results in many schools repeatedly reveal that some children who are given the "best curriculum" chosen by a school end up in grade five with terrible skill levels. These children have had four years of wasted instruction in math, and their skills remain stuck at grade-one level. For four years, nobody was doing adequate monitoring of the children, and nobody was creating new tricky mix adjustments to try out.

## Treatment Resisters in Psychotherapy

Regardless of whether a regimen of treatment is drug-based, therapy only, or drug and therapy, in nearly every research project a subgroup of patients treated for anxiety and/or depression show up who make minimal progress or regress. The label "treatment resister" is applied, but typically it remains unclear why they have done poorly.

## Deaf Children's Literacy Skills

For many decades in the USA, the preferred "best strategy" for reading instruction was to assign the deaf children to special residential schools. In these schools, it was expected that the children would benefit by being with similar peers and by receiving instruction that took their specific hearing impairments into account. Results showed that this overall strategy was far too simplistic. By grades six to eight, for most deaf children, their literacy was too low to support good progress in academic areas and too poor to support entry into college programs. In our own initial research on this topic, we found many deaf students at grades five and six who could not read *any* simple phrase or sentence. Again, we see a case where these students had encountered four to five years of reading and writing instruction that was totally wasted.

## Head Start Children Trying to Catch Up to Peers

When three-year-olds enter a Head Start program, their communication skills usually are far behind the eight ball. For example, compared with their non-Head Start peers, they average about 10,000 fewer words in their vocabulary.

A year with Head Start teachers and aides often is considered a "best" option, one that is expected to be highly stimulating and sufficient to close the gap in language skills. Alas, this prediction is seldom even assessed—so confident are program designers and administrators in the program's efficacy. And our own research based at Penn State University has very bad news indeed. Being in Head Start from age three to four does not raise the average (poor) skills of these children at all compared with the skills of children who had no Head Start experience between three and four years. In short, for language goals, the Head Start investment was a total waste of time for both the children and the teachers!

## Professors Trying to Predict Which New Graduate Students Will Have Outstanding Careers

I was fortunate enough to have the experience of graduate education in psychology at Yale University. They even awarded me a PhD!

More than that, Yale was located close to Long Island Sound and some excellent sailing waters. Further, two friends introduced me to both the cruising aspects and racing aspects of sailing. And so, beyond whatever I learned about psychology research and thinking, I also gained incidental sailing skills and motivation and ocean awareness that have led to a terrific series of life events.

Another incidental that I learned while at Yale was about some of their own internal research on admitting undergraduates like me into their PhD program, as well as failing to admit others. The faculty opinion had long been that one of the "best predictors" of later success in this field was high ratings from personal interviews with the faculty held before admission was offered or denied. By looking decades after application, it is possible to see who achieves high success in psychology. When a few faculty members finally looked at such data, there was a shocking finding. High interview rankings for applicants did not predict long-term career achievement. In fact, overall, interview ranking levels were weakly but negatively related to career achievements and significant breakthroughs. Kind of makes you curious, doesn't it, about other hiring/admission decisions where heavy emphasis is placed upon what an interviewer thinks about an interviewee?

What we see here, again, is that what seems to be the "best option"—even if highly qualified persons have chosen it—will often lead a group astray. If confidence in that "best option" is high, this will lead to a reluctance to do valid and timely evaluations.

## Huge Egos and Over-Confident Huge Investments

It is not reassuring to the ego of anyone in a responsible position to collect data and find out that for years you have been wrong in your judgment of what will work well. As a result, over-simplistic attempted solutions to an issue will often be protected from scrutiny. Moreover, that tendency is amplified when the lead persons for a project are rich, famous, and authoritarian in their decision-making—they will love having their name splashed around for the project, and the bigger and longer their investments, the more they will resist needed abandonment of a failed mix of strategies.

## Two Themes Cutting Across All Chapters

The scope of possible advances will be underestimated when dynamic tricky re-mixing is limited.

Even a modest increase in understanding the dynamic complexities of situations and challenges opens the way to powerful new tricky mix strategies.

Chapter 9

# SURPRISING RECOVERIES
# AND LIFE-PATH SHIFTS

U nanticipated and extreme changes in circumstances sometimes set in motion re-mixing that otherwise would be highly unlikely. The first case of this we will examine is one where profound positive effects were produced both for a clinical patient and for an impromptu therapist brought into the situation.

## Rehab Innovation Transforms Both Patient and Doctor

Paul Bach-y-rita had a well-established medical career in place when his father suffered a severely debilitating stroke. One day, his father was enjoying his own career as a poet and professor. He loved his work and was enjoying considerable success and recognition in the Catalan region of Spain. The following day, a stroke hit and he could no longer talk, write, walk, or even crawl. Even months later, the rehab that was given to his father seemed to have no discernible effects on these life-limiting symptoms.

So, Paul and his brother plunged into their own made-up rehab exercises based upon Paul's sense that his father now needed to relearn the kinds of skills that infants and toddlers gradually learn. Paul would assist his father in crawling, reaching, and other simple motor tasks. For instance, after some weeks, crawling exercises with constant therapist support gave way to crawling with a wall on one side as support, to crawling with no support at all, to the first supported movements toward walking.

Prior rehab procedures did not produce these simple advances in motor skill and mobility, and it would have been easy to simply conclude that there was too much brain tissue destroyed for new learning within the remaining intact tissue. His father, to many, seemed *stuck*. Paul's more complex analysis was essentially that by radically re-mixing conditions, gradual change would be feasible. Add in positive expectations of change, loving and involved sons as therapists, whatever daily exercises current levels of skill would allow, and then re-mixing again as simple advances were made to introduce new challenges, leading to ever-improving skills and changed expectations.

Fast forward two years, and the elder Bach-y-rita, professor and famous Catalan poet, was again teaching, writing, and enjoying the fruits of his long years of excellence and productivity. The recovery was astonishing relative to other stroke victims with similar initial symptoms. An almost total re-mix of his rehab had produced a life-changing and complete recovery! Note that an essential part was the continual monitoring and adjustment process—the complexity-based dynamic re-mixes—that revised and updated the rehab procedures in order to keep achieving positive tricky dynamic mixes and successive advances mentally and physically.

The clarity and surprise of what he had achieved for his father had an additional effect on Dr. Paul Bach-y-rita. He recognized the tremendous importance of brain plasticity that was proven in his father's case, and the import for countless other cases where rehabs may be underestimating the capacity of stroke victim's remaining intact brain tissues and circuitry. So, Paul made a life-changing decision for himself. He retrained and devoted the remainder of his career to understanding the brain and body better and helping to revolutionize rehab approaches.

This cascade of accumulating positive changes for Paul and his father provide the answers to the "poet" dynamic puzzles in Chapter 5.

## From Strange Kid Lacking Basic Skills to Class Star

Imagine that you are a student in a very small school, completely failing in reading and in all other subjects but passed along each year with your classmates until you reach high school. What changes in self, if any, could set your learning on fire?

This was the situation in Australia for Donna Williams, a child who entered school with severe autism and a preference for imaginary playmates versus real social partners. Her autism symptoms persisted into high school, along with her low engagement with school subjects. Yet, a few short years later, she achieved the top performance in school-wide mandatory achievement tests!

At the heart of this phenomena was a fascinating new mix of learning conditions in one drama class. Her drama teacher bypassed Donna's skill limitations and inserted her in on-stage theater roles with complex lines. Because Donna could not read the script, her teacher read the lines to her and soon she knew the lines by heart. She was a capable actress! She was a cooperative social agent in the play's complex unfolding. She also was recognized for these skills. And, in consequence, her self-concept shifted radically. Now, she knew she could learn complex material, so she strongly expected she could be self-efficacious in future classes and studies. Moreover, she used the key insight that reading and writing opened up new worlds to her and learned to read at high levels of competence.

In tricky mix dynamic terms, there were strong convergences of expectation, curiosity, persistence, and positive emotion. Things snowballed and, as she learned more and more, the long-term memories were enriched, reading fluency raced upward, her confidence kept ratcheting up, and all the advances fed into better dynamics of learning.

This set of rapid gains in skills, confidence, and positive self-concept extended beyond successful completion of high school into her eventual career. Donna became a well-known author whose autobiographies have inspired many other individuals on the autism spectrum and those who are devoted to helping them.

## When a Blind Man Triggers New Vision in a Sighted Man

You will recall the dynamic puzzle from Chapter 5. Why would a one-time occasion of noticing a blind man along a sidewalk trigger change?

The answer lies in a rapid re-mixing in the sighted man, my friend John. John let himself be affected by this so that new insights and deep emotions dynamically occurred together. As if smacked on the side of his head, John realized that, in multiple ways, he was "blind" to his own inner nature, to dissatisfaction in the career path he was on, and to the dysfunction in other recent choices in his life.

Soon, there was a snowball of dynamic changes emerging together. John shifted his career path from engineering to psychology. Along this new path, he met his long-time partner and wife, Carolyn, began a satisfying path as a parent, and acquired tremendous skills in counseling and teaching. All these positive changes were set in motion by the strong dynamic tricky mix of new evaluations and new choices, all flowing from truly noticing one blind man proceeding smoothly along a physical path despite the challenges sightlessness raises.

## Learn to Treat Many Forms of Bacteria as "Honored Guests" in Your Body: Experience Cascades of Well-Being from a Newly Achieved Balance

From the outset, this book has warned against simplistic, "silver bullet" attempts to solve a personal or societal crisis. Yet, some of the most egregious insults to our bodies come from simplistic medical attempts to "cure" discomfort in our gut or elsewhere through the strategy of strong doses of an antibiotic.

In contrast, rapid declines in inflammation and discomfort and rapid increases in mood, energy, and overall wellness may proceed from the introduction of better nutrition and probiotic supplements. Rather than totally wiping out a strain of bacteria that has become too abundant, this approach allows that strain to re-establish a healthy relationship with a new balance of bacterial and virus strains. Better immune systems, better neurotransmitters, and more effective nutrition flow are some of the benefits. All these benefits were reaped by Professor Dietert when he made a major shift to probiotics

His case, earlier presented as Dynamic Puzzle #1 for you to solve, dear reader, is now ripe for revisiting.

You were asked to predict whether a re-mix of conditions might lead the professor out of his persistent gut, bowel, and sinus problems. Let us first recap, then update.

He took many antacid prescriptions in an attempt to control his irritable bowel syndrome. Further, he developed persistent sinus symptoms and would take a round of antibiotics that seemed to help until he stopped the antibiotics and symptoms recurred. This repetitive cycle just kept going. Even worse, additional infections and periods of low energy were frequent, and gut, bowel, and sinus problems recurred. Dynamic Systems were certainly at work, but with cascading *negative* outcomes.

Antacids and antibiotics as simple "silver bullets" clearly were not solving this professor's problems. A breakthrough was triggered, as so many are, partly through a lucky change in context. A week spent in Germany with colleagues changed his diet for that week toward local, fresh, organic menus. Soon afterward, he felt such physical improvements that he knew it was crucial for him to re-mix his nutritional plans. That replanning incorporated multiple powerful probiotics and a healthier range of foods which then led to all the cascades of positive health changes described above. It also led him to share his kind of re-mixing with many others who also made crucial life recoveries. Professor Dietert's book on this topic is *The Human Superorganism: How the Microbiome Is Revolutionizing the Pursuit of a Healthy Life.*

## Lessons from the Planet's Greatest Natural Predator

*Top Predator, the Fierce Great Horned Owl.*

The fantastic tricky mix of features which makes this "tiger of the skies" nature's most efficient predator includes some you can notice in this close-up view. His eyes are large, especially rich in night receptors, and placed on a head that swivels to cover the full three-hundred-sixty degrees around the bird. Feathers around the eyes channel sounds to the ears. And these ears, located at slightly different heights on the head, further make location-finding precise. The beak and talons are large and powerful. His wings are powerful as well, and built wide and short enough for skillful maneuvers around any wooded area. In addition, he is equally adept at locating and seizing prey of all sizes, from tiny voles to rabbits to geese.

My closest encounter with this "tiger" was right here at home on our farm. One surprised me by taking off from the ground just a few feet away from me, flying directly over my head. Wow! The rush of wind from his wings was strong and, from my viewpoint, those wings were enormous and threatening. Fortunately, he bore me no ill and flew on his way with his fresh catch dangling from his golden talons—a large, plump rat!

Yet, even for this fierce owl, dynamic conditions sometimes mix to create temporary *stuck* episodes. Below, an image captures another great horned owl who has been grounded by an unusual *mix* of conditions. This owl—like an overloaded airplane—is, first of all, temporarily too heavy for flight after a large meal and has chosen to rest here beside our stream. Further, a heavy rainstorm has moved in and our owl is so wet he appears a bit out of focus in the camera. The soaked feathers make him even heavier, not ready for flight. He appears disgruntled, but after a while, the various prey in his belly will be digested, he will defecate to lessen his weight, the rain will pass, and he will shake his flight feathers free of water and soar once again.

*A Surprising Negative Tricky Mix
Has Grounded This Owl.*

Chapter 10

# EXTREMELY RAPID CHANGES
# & THE RARE EVENT PRINCIPLE

S o far, you have seen many exemplars of how dynamic tricky mixes work. People, projects, and organizations get *stuck* when numerous well-meaning efforts fail to bring together a full set of learning/change/growth conditions. In contrast, from the same dynamic processes, we have seen how remarkable change over weeks or months has been triggered by the innovative creation of a new mix of supportive yet challenging conditions.

Here, we will examine some cases where remarkable change and reorganization has been created within just minutes or hours or, at most, one week. We will also bring in the concept of rare events as part of the overall landscape of dynamic mixes.

## Phantom Human Limbs

Phantom limb experiences have been documented in a myriad of cases involving loss of arms, legs, or other body parts. The essence of this strange experience is that a person with limb loss "feels" that the limb is still there as a compelling ghost or phantom. Such phantom limbs, despite obviously having no nerve connections to the brain, sometimes experience excruciating cold, heat, itchiness, or pain. Pain for some patients is so intense and frequent that it distracts from most activities, leads to depression, and even leads to suicidal thoughts. The afflicted person knows that in an objective sense the limb is truly gone, yet the subjective experience of a limb that is "still there" and feels pain is powerful and undeniable. For many years, these symptoms of phantom limb pain were untreatable.

Then, along comes a neuroscientist named V. S. Ramachandran. He argued that the brain was dynamically tricking itself by inappropriately mixing long-standing, highly learned memories and expectations of what a person would feel in their limb with the absence of any actual input to or from the limb. So why not try to "trick" the brain back to a more realistic, functional, and pain-free model of what's going on?

His lab, and many others since, set up a "smoke and mirrors" magic-like trick, but without the smoke. Using a simple box with mirrors in the middle and two cut-out holes where an arm could be inserted on each side of the mirror, the patient with phantom limb pain was instructed to insert their real arm in one side and their phantom limb in the other. Looking down into the mirror, the patient now had visual input that seemed to show a miracle—as they moved their intact fingers on their real hand, it looked like the fingers on the phantom hand were moving. This forced a puzzle on the brain that was new—how to make sense of sensory input coming from the intact hand with the visual perception of the phantom hand moving, yet without sensory input arriving from that hand?

For many patients, this kind of mirror-box experience across just forty to eighty minutes led to a great success: the phantom limb and its persistent pain both disappeared. Full relief from the pain!

This brief, experience-triggering change we will label a rare event. Rare events are just one part of the spectrum of dynamic mixes that lead to change, but they are both fascinating and important. When a powerful set of learning/change-causing conditions converge that include all relevant components and positive levels of each, then one, five, or ten brief events trigger substantial, significant change. We will have a tour in this book of many such examples where the same basic processes of Dynamic Systems changes lead to sudden leaps forward—but only when the converging conditions are positive and, most often, novel.

To recap, phantom limbs and phantom pains are constructed by dynamic processes within the brain and body that are closely similar to more functional and realistic models of our bodies and the world it interacts with. The brief experience of looking into the mirror box allows the brain an opportunity to dynamically reorganize the way that prior memories and expectations are integrated with current experiences. Re-mixing dynamically thus can create a calmer and more objective body awareness. This new integrative process, when successful, leads to the rejection of any apparent pain or feeling coming from a long-lost limb as dysfunctional and erroneous.

Here is the rare event principle that applies here and widely across the experiences of children, adults, and organizations.

> Substantial learning, change, and recovery can be driven by one or a small handful of infrequent events. Rare events can trigger needed advances.

## Quicksilver Minds of Two-Year-Olds

Instances of incredibly fast, substantial learning are not restricted to adults. Indeed, there are many more well-documented illustrations from the lives of children.

As we will see from rigorous research with children, rare-event learning in children under three years of age reveals a wealth of insights into the cognitive mechanisms of these youngsters.

Our first exploration involves specific language advances by children from one-and-a-half-to-three years of age that the majority of experts had predicted would be impossible. In English and many other languages, the "active" form of a sentence might involve an actor, their actions, and the object of the action—expressed in that sequence. Thus: *the little brown cat chased the little white cat*. In "passive" structure, the actor/object reverses, and we see: *the little white cat was chased by the little brown cat*. Multiple claims were staked that the passive form was too complicated for the mind of a child three years or under to grasp. This is a strong claim, and one that was easily disproved in our lab.

Recall that a key theme is that just because the usual circumstances or "learning conditions" do not support progress of certain skills, those skills may rapidly progress if a radical re-mix of learning conditions occurs. We tried that for passive sentences with children two-and-a-half-to-three years of age. Samples of what children knew about passives revealed that, before our re-mix experiment, the children studied had no passive sentence production or comprehension. Good! Now we could answer yes or no to the question of whether our re-mixed conditions could trigger passive acquisition. Results cried yes! What had been claimed as impossible was very possible indeed, and for many children within just two forty-minute sessions of re-mix conversations.

So, what were the core elements of our effective new learning mix?

1. We had a fun conversation with each child based upon their interests and their own expressed sentences.

2. In our replies to a child's sentence, we mixed in passive structures as part of our responsive and engaged conversation. So, if the child said, "That dinosaur hit the hippo," we might say, "Yep, that hippo was hit by the dinosaur." Our reply reflects a recasting-re-mix strategy. Each recast reply *maintains the child's basic meaning but recasts that meaning into a new sentence structure which holds high potential for the child to learn something new about language/syntax from the recast.*

3. Engagement and motivation were kept high by way of the teacher/therapist showing positive interest, emotion, and enjoyment of the ongoing interaction with the child.

4. In combination, this dynamic re-mix of learning conditions guaranteed many passive examples could be learned because they were given unusually high attention, analysis, and salience in the child's ongoing processing of the sentences and their relation to context.

Make no mistake. The rapid learning of passive language structures was surprising because so many had predicted it could not occur given the assumed brain capacities of children less than four years of age. Moreover, because the experiment created the conditions specified above, we knew what we usually don't know—the actual interactions between a child and an adult that causally supported advances in language. These experimental results demonstrate three key conclusions:

1. Most children three and under have powerful cognitive capacities sufficient to deal with passives.

2. Many such children also fit with the rare event principle as they launched their new use of passives within one-to-three hours of the "recasting re-mix."

3. Nearly all of the children learned at rates that were impressive, showing acquisition either at the Rare Event level or a bit later, by four-to-twelve hours of interaction.

Once again, the principle: substantial learning, change, and recovery can be driven by one or a small handful of infrequent events. Rare Events can trigger important advances or shifts.

This is termed a "principle" because it can apply to a broad range of skills, concepts, and contexts.

As one further illustration of this point, experimental work with children as young as fifteen months has demonstrated Rare Event learning of new concepts and their names. What is particularly informative about this research is that the input learning opportunities are strictly controlled. Because neither you nor a fifteen-to-eighteen-month-old could initially know what a *Fiffin* or *Bandock* might be, we can study you like a blank slate. So, in our lab, we made up such words and the objects they denote and made sure that the only learning opportunities happened when the children visited the lab. And to test the "rare event" idea, we made sure that, for some of the target names/objects, only one object illustrated the target. Thus, we could find out—unlike most name-learning situations in the homes of toddlers—if encountering just one object and its name every two weeks was all that a child needed for learning. The answer: yes, most children did learn from such a "rare" or "infrequent" exposure. They learned names like *Fiffin* and *Tallrik* readily from extremely infrequent input. "Quicksilver minds" accordingly seems to fit these young children at one-and-a-half-to-three years of age when it comes to learning vocabulary, grammar, and many other fundamental areas of knowledge. Modern cultural complexities would, without a doubt, be impossible if children at these early ages were not such fantastic learners.

# Six Surprisingly Positive or Forgiving Lightning Storms

Without a doubt, a lightning strike can often be intensely destructive. Dynamics of lightning events are well understood for the most part. A negative charge at the lower part of a cloud can be so great that it can repel electrons on the ground up to five miles away. When the electrons on the ground are repelled, the ground becomes positively charged, now attracting the negatively charges electrons at the bottom of the cloud.

Ready, set, strike! Now the bottom of the cloud and the ground are like giant magnets being pulled toward one another. The negatively charged particles of the cloud race toward the ground at a speed approximating 300,000 km/hour (about 186,000 mph). When they hit the ground, thereby neutralizing the difference in charge, we see a brilliant flash of light that delivers 300,000 volts and causes the temperature of the air it passes through to reach up to 50,000 degrees Fahrenheit.

When an electrical shock of 300,000 volts is delivered to the body and brain, this can cause the heart to stop and can also cause significant brain damage, usually in the form of long-term memory loss. The hot air accompanying lightning causes third-degree burns, and the discharge and heat together can cause blood vessels to burst, leaving lightning bolt-shaped burn marks. Occasionally, the brain is instantly blown up, resulting in sudden death.

Dynamic Systems tell us, however, that a powerful strike of lightning—or sequence of strikes—will have dramatically different effects depending upon the mix of local conditions at the time of the strikes. The following examples show some positive outcomes following lighting strikes that created either no negative effects or fairly transient ones.

## Tennis Misjudgment: When Lightning Takes Over a Tennis Serve in Texas

In Chapter 5, we encountered an engineer, Jack, who posed the dynamic puzzle of what follows when you are knocked down by lightning. Jack is a great friend and colleague of mine, so I have explored this event in detail with him. Here we pick up more of his own words about the event:

"A storm was brewing in the distance with lightning dancing in the clouds headed in our direction. Our informal rule was to play until the rain soaked the court. It was a bad rule.

"As I raised my arm to serve, *crack, bam,* a bolt hit me on the head. As I fell, I heard a sharp snap from my tennis shoe to the asphalt surface. Sprawled out on the court with rain pelting my head, I struggled to regain consciousness from what appeared to be a very bad dream. I glanced across the court to see my partner scrambling to his feet. This scene was not a dream. We had been hit by lightning!

"Tony and I checked each other out. As best we could tell, our bodies appeared to have no physical damage. We speculated the electrical pulse must have traveled down the path of least resistance, our body sweat. After a few minutes, Tony decided to leave. I was still not fine, feeling dizzy and brain-fogged.

"After the storm passed, I decided to drive home even though I was still in a fog state. I told my wife, Sheila, and our kids, Jennifer and Bryan, what happened to me. They reassured me that I appeared to be alright in their eyes. I assumed my brain would clear out the fog in a few hours or, at most, overnight. It did not.

"The next day I remained in shock. That afternoon, driving home on a Houston freeway, a strong thought flashed through my mind: "I could simply turn the steering wheel and kill myself by smashing into a concrete abutment." I had never experienced such an impulse. Something was very wrong with me. Suicide? Could it have been caused by the lightning strike? I needed to find out, and quickly, before my foggy brain malfunctioned again and left me dead on the highway at age forty-two.

Well, without a doubt, Jack's lightning strike produced very rapid dynamic changes. What happened for Jack over the weeks, months, and years following the event proved to be dependent upon the new paths he explored and his exploration of a series of new tricky mixes. In later chapters, we follow up on Jack.

## Lightning Creates Glass at Farmyard

Returning from a trip abroad to the farm country in central Pennsylvania, we discovered more about the power of lightning. Like a mystery novel, at our farm and farms nearby, there were multiple clues scattered around that lightning had been at play while we were gone.

One tree near a neighbor's front porch was heated so hot that all the sap and moisture disappeared, leaving a hollow, light tree trunk more like balsa wood than its natural maple . A Volkswagen Beetle was hit, but left unharmed except for the electrical chip that was the brain of the car's systems. At our farm, two maples were charred and modest splitting of limbs occurred.

Beyond these clues, the most intriguing was a lightning strike that altered a wooden play structure and the sand beneath it. My young daughter loved this structure as it had a ladder leading to a deck on the second level beneath a slanted roof. On the deck, it was fun to pretend to be sailing a boat by steering the fake boat wheel or, alternately, weave out all sorts of other house/family fantasies. Beneath the deck was a rain-sheltered sandbox area cluttered with a few metal spoons, plastic buckets, and small toys. All this was changed by lightning. The whole roughly eight-by-eight-foot structure was twisted beyond repair and scooted about ten feet away from its usual place. Toys and buckets were thrown even further.

But the most remarkable transformation was where the sand play area lay. There, just below the sandy surface, was an alchemist-like conversion of ordinary-colored sand into yards of strange green glass. I had never seen nor heard of such an event, but yet another neighbor instantly remarked: "Oh, just like in the movie *Sweet Home Alabama*." Indeed. In that story, a sculptor intentionally attracts lightning by planting metal rods into beach sand. Later, he checks below the surface for newly created glass after a thunderstorm passes. The glass thus created has been termed "fulgurite."

Our down-home Pennsylvania fulgurite glass was created in sheets about a quarter-inch thick with a pretty light green cast, and so light to the touch that it was easy to break into smaller pieces the way you can break most cookies up in your hands. For us, this encounter definitely was a rare event showing us, for the first time, the power of extreme heat when it was generated by the particular pattern of lightning that had drifted through our local farm area one night.

## Open-Ocean Sailing: No Way to Evade Lightning

On an eleven-day open-ocean sailing trip from Tortola into the Bermuda Triangle, and then on to Virginia, myself and four others were often surrounded by intriguing variations of water and light. Near the completion of our journey, conditions turned more threatening. We could see thunderstorms on the horizon to the west as we continued to advance strongly along a fast point of sail. Soon, we were in the first of six thunderstorms. As the intensity of the rain and wind increased, Lars, our captain, and I were dressed head to toe in rain pants and jackets, driving the boat hard through the waves. Others rested below out of the tempest. Thunder and lightning moved in close to our soaked sailboat. It was hard to distinguish in all the movement and shifting of light and water where the ocean and sky began and where we fit in between. Nevertheless, the boat was holding up well and there was exhilaration in our rapidly slicing through the wall of falling rain and wild waves. Lars and I chatted away over the roar, leaning our heads in close together and both holding on to the aluminum wheel that steers the craft.

Suddenly, we heard a loud hiss and snap right on top of us. We both glowed and crackled, and the boundaries of our bodies, soaked rain gear, and the air around were blurred. There was a count of six and then silence. We looked at each other and realized that lightning had struck and passed right around us and down the sides of the boat into the ocean. Also, we were amazed that we were unharmed.

We beamed at each other. Whew! Vision, hearing, touch, movement, speech, tears, laughter . . . everything still worked! What a close call, and how fantastic just to be alive and still sailing a boat in a roaring storm, still looking forward to what the next turn of weather and events would bring. We might have been toast. We might have been blackened tuna. Instead, life continued with a new vibrancy.

Needless to say, I was in no position to record a picture of the lightning striking passing over the boat and our bodies. What sticks in my memory is the remarkable, dynamic emergence of the whole experience, from the complex of waves, the moving storm, the charging boat, and the distant thunder and lightning. And, of course, that new vibrancy which emerged after the strike of lightning. This was a change in self-concept that has—right up to the present moment—richly fed into my desire to lead a life helpful to others and to all of nature's creatures, and to do so with a renewed sense of purpose and a passion for innovation and creativity.

## Music Passion Sparked by Lightning and Other Curious Brain Transformations

When lightning or other remarkably novel events impact our brains, the outcomes vary extensively across individuals and depend, in part, upon the already ongoing dynamic patterns in our brains.

The late, notable, and highly curious neurologist Oliver Sacks has provided some brief case examples which deserve attention within our dynamic tricky mix perspective. His 2007 book *Musicophilia* is a treasure trove of such cases.

First, "Lightning and a Surgeon," the case of Tony Cicoria, MD, a surgeon who was struck by lightning at the age of forty-two in Albany, New York. It happened in 1994 when people still needed to use public telephones when away from home. Tony had just finished a phone call to his mom from a metal-framed phonebooth when he was struck. A flash emerged from the phone and knocked him down. Then he had an out-of-body, near-death experience before a stranger revived him.

He suffered relatively minor impairments, burns on his face and foot, plus some problems remembering names for friends, diseases, and the like for about a month. He returned to work and the memory issues soon resolved.

But there was a very intriguing further transformation. About three days after the event, Tony found he now had a burning desire to listen to piano music. He did this, and his passion for music deepened further. By lucky chance, for the first time, a piano was available in his house because someone needed to store it. So now he pursued learning to play Chopin and other music as well as original music that emerged spontaneously in his head. Within Dynamic Systems, it seems safe to say that the lightning was a strong perturbation which caused new connectivity patterns within Tony's brain that led to a new sense of identity as someone who loves and obsesses over music. As part of this new identity, he also developed new, deeper spiritual feelings. These changes constituted a far-reaching tricky re-mixing of self-concept, core behaviors, and brain connectivity dynamics. Remarkably, all of these were integrated so that Tony's identity and competence as a surgeon also persisted.

Pronounced shifts in identity, personality, and pursuits are sometimes introduced by a variety of strong, challenging perturbations to the human brain. For one female research chemist, a brain tumor was removed in order to reduce seizures. Her brain, as in all such cases, had to make dynamic adjustments to those tissues still intact after the surgery. Fortunately for her, as in the case of Tony's lightning strike, after a few months, her professional skills were fully available and she resumed her career. One curious side effect was that her personality shifted from being a reserved person to being far happier and warmer, as well as being a more empathic, socially outgoing, and highly emotionally expressive person. The second change was, like Tony, around music. Now, for the first time in her life, she craved classical music, listening whenever she could on radios, at concerts, and so on. She loved the music and responded with high emotion—with joy.

In dynamic tricky mix terms, it is important to recognize, the convergence of multiple new and complex conditions in each of these cases. First, brain dynamics shifted toward more fluid and joyous emotionality. Second, interest in music skyrocketed. Third, time spent in new musical activities also flourished. Fourth, these new activities led to new neural representations of the new musical experiences, including both musical structures and sensory experiences. Fifth, over time, the richness of all of these conditions was growing, creating more and more pleasurable anticipation, engagement, and expanded musical experiences. Sixth, engagement and rates of acquiring new skills in music also improved.

Sacks, among others, also documented cases of surprising positive outcomes of an insult to the brain from a tumor or car crash in the *visual* realm. New forms of expression in painting may flourish after such insults. In combination with the evidence that some individuals, as illustrated above, leap forward in musical domains, Dynamic Systems theories fit well. Whether the insult arises from lightning strikes or other unpredictable events with common basic dynamic processes, the brain, in fortunate cases, establishes fundamentally new bridges and new integrations between areas concerned with emotion, goals and planning, auditory domains, visual domains, and motor execution.

## Trapped and Entertained by Thunderstorms: It Was a Dark and Stormy Afternoon

It was a dark and stormy afternoon. At our farm, Eagle Spirits Farm, my daughter Leilani and I were inside the large, Amish-style barn talking to the goats. Occasionally, barn swallows swooped by on their way to their nests or outside through the wide-open barn doors. We were relaxed, feeling peaceful and grounded.

Then the sky outside turned dark. Soon, the dark clouds were releasing torrents of rain. Lightning struck between the barn and the stream in front. Very close! And the intense thunder clap followed immediately. The smell of ozone saturated the air.

Leilani and I flowed with the situation, feeling safe and dry but intensely stimulated in all senses from the flashing light, loud booms, and moist "burned" air. Further, the swallows were now staying in the barn and closely fluttering just inches away from us, chattering to their chicks in their nests.

We were caught up in the storm raging outside and the swallow "storm" floating around us. We counted seven different thunderstorms as they all released torrents of rain and cascades of lightning. Our house was just a thirty-second dash away, but we felt the storms made that an unsafe idea. Finally, with just the slightest slowing of the storms, we ran and made it to the house. There, we felt safe again, and, at the same time, grateful for the exhilaration and the wonderful lasting memories to treasure.

## Great Balls of Fire

Twice in our family, fiery, scary balls of lightning have entered a house.

When my wife Kathy was a young kid in central Pennsylvania, she was at home with her siblings and her mother. Suddenly, a glowing ball of lightning rolled around the floor close by. It then blew out the TV set. The positive side of the situation was that all the family members were perfectly unharmed.

In the case of my parents in Kansas, living in "thunderstorm alley," they were both at home and were untouched when ball lightning formed. The noisy, intense ball of fire rolled right along the power line and into the house as they watched. It soon landed in the living room, blew out the TV, and scorched the surrounding area, setting off a smoke storm.

These instances are both rare events, and the remarkable dynamics that created them could so easily have led to injury or death. The particulars of the situations led to a more fortunate mix of conditions and spared those who were close to the lightning discharges.

*Dynamic Lightning on the Move to Human Structures.*

# A Very *Stuck* Nine-Year-Old Gets Rapidly Un*stuck*

When all else seems to have been tried with a child on the autism spectrum with minimal success, a radical re-mix of conditions may open a door to a workable path.

For nine-year-old Ted, a boy with excellent hearing but almost no spoken language, John Bonvillian and I tried something that seemed very "out of the box" to most at that time. Namely, we induced his parents and teachers to interact with Ted in sign language. Ted, you may recall, was presented as a puzzle to help solve in the Preface.

This move to sign language proved to be a step in the right direction. Ted was able to engage his learning mechanisms effectively to start comprehending and producing signed words and phrases in the very first weeks of sign language conversation.

His further progress sparked enough sign language to support—for the first time—reading and educational engagement elsewhere. The key point is that one radical change was the rare event that shifted Ted to a more positive trajectory. Subsequent researchers followed up on this approach and documented that, for some other children with autism, a shift to sign language as a visual mode of communication served to move children out of *stuck* position developmentally, triggering remarkable new achievements in their language and social behavior.

Chapter 11

# QUANTUM SHIFTS: QUANTITATIVE UNDERPINNINGS OF MAJOR QUALITATIVE SHIFTS IN SYSTEM PERFORMANCE

Introductory note: We use the phrase "quantum shifts" as a metaphor for substantial and often sudden shifts to new modes/levels of behaving, sensing, thinking, exploring, implementing plans, and "living experimentally."

On the surface, some breakthroughs are so qualitatively evident that entering into any quantitative analysis or discussion may seem pointless. If a person is non-talking and, when given L-Dopa, springs into talking and ordinary movement, then that breakthrough might seem to require no further analysis or explanation.

We will argue that, even for such cases, it is valuable to consider quantitative aspects of which factors and what convergences are at play.

In this chapter, we will introduce ways of quantifying the strengths of tricky mixes and show that some solutions to the dynamic puzzles raised early in this book are well-described within these quantitative frameworks.

We will argue that no prior accounts of human motivations give due justice to the variability and complexity of how motivation and multiple other factors dynamically converge. This is true for the efforts and plans of both individuals and groups. The dynamic tricky mix approach recognizes that dozens of "motives" have been explored by prior theorists. Among such motives are achievement, curiosity, self-efficacy, mastery, power, and altruism. We argue that behaviors are usually driven by emergent dynamic convergences incorporating multiple social-interactive, situational, emotional, and motivational factors. Beyond usual motivational accounts, our approach emphasizes converging forces toward an objective.

## Qualitative Levels of a Dynamic Tricky Mix Approach One: Probability Specifications Within Five Modes

Here we explore how likely it is—given a specified mix of factors—that desired progress toward a goal will occur. Such progress may take many forms. For example, progress might be toward new learning, a creative breakthrough in solving a problem or issue, a positive change in the trajectory of a group, or a new level of executing skills already in one's repertoire.

Five qualitative categories of change/progress are employed to organize accounts of what factors are at play, the strengths of each factor, and the probability of progress that is created by the dynamic convergence of the factors. Each of these categories is a mode that will tend to persist until someone or something nudges the existing dynamic mixes occurring in that mode.

> *Stuck "Doggies"*
> *Moving*
> *Improving*
> *Flowing*
> *Soaring*

### Stuck "Doggies"

When no progress is being made, it is essential to analyze dynamically what mixes of factors are occurring. In far too many instances, zero progress occurs despite high optimism that planners (whether individuals or groups) have made excellent choices and already put in place necessary and sufficient conditions for success. What could possibly go wrong? Well . . . plenty!

Faced with well-documented failures of individuals or groups to learn, change dysfunctional behavior, take constructive joint action, perform well even after extensive training, or meet other desired criteria of progress, what is the response of many planners? The response is to blame individuals for lacking sufficient motivation, ethics, or intelligence for appreciating and benefitting from the care, expense, opportunities, and guidelines provided by the planners!

Case Example: conditions influencing progress toward a drawing of a 3D house by a young child who had previously never drawn such a house. We will use this case to see a mix of potential factors for the STUCK category first then for each remaining qualitative category.

*If the only two factors (one dyad of factors) present are if the child has (1) drawing materials and (2) a 3D house drawing to refer to when drawing their own, then the child will remain stuck at his/her current level of drawing—that is, a very simple 2D house based on a small, simple art tool kit.*

A child who has only a small, simple art tool kit will not only draw very simple houses, but will create similarly limited renditions of any subject. Here is a boat created by a child four years of age with a small set of artistic techniques. The piece is lively and colorful, but based on a few repeated techniques.

*Child's Depiction of Sailing Boats.*

## Moving

Events taking place in this category/mode of progress lead to inconsistent gains. New mixes of conditions have emerged that get the individual or group past *stuck*. Still, the contributing factors themselves are fragile and the dynamic convergences among the factors are also fragile.

Probabilities of some slow progress will fluctuate between 5% and 40%.

If, in the past, an individual or group had been *stuck* for a while, it will be refreshing to see some gains occurring. Moreover, as time goes on, there will be an accumulation of episodes of better skill execution or distinct learning of new skills/knowledge. This, in turn, may lead to shifts in positive expectations, self-confidence, focused attention, and strategy selection. In consequence, a stronger set of factors may emerge that lead to a "quantum leap" toward better dynamic mixes and entry into one of the higher progress modes—*Improving, Flowing,* or *Soaring.*

To get a child moving, here is a mix that has been shown to support a small amount of progress.

1. Learner sees a friend try
2. Learner sees a friend make tiny progress
3. An adult encourages Learner to try
4. An adult gives demonstration of one component technique of 3D drawing

## Improving Level of Progress

1. An adult encourages Learner to try
2. An adult gives demonstration of one component technique of 3D drawing
3. An adult with modest artistic skills shows how they go from a blank page to a complete 3D drawing of a house
4. This artist refrains from telling the child what to do
5. The anxiety is low
6. The child experiments with new house drawings

With these six factors (three dyads of factors) present and converging, probabilities of some modest progress will fluctuate between 50% and 75%.

## Flowing

1. An adult who is well-liked by the Learner directly encourages the Learner's drawing
2. An adult who is a highly skilled artist and an emotionally significant partner to the Learner fluently makes their own complex, distinctive 3D drawings right beside the Learner
3. The anxiety is low
4. The artist makes no direct commands or instructions to the Learner and instead allows the Learner to choose what aspects of the artists' creations are of interest
5. The child Learner sees their own drawings improving
6. Learner "shoots" the personal sacred cow of "I lack artistic potential"
7. Learner has seen new live drawings in progress and experimented in response
8. The child Learner receives verbal rewards for their interest and engagement in drawing

When all the conditions of this mix converge, then high rates of progress are likely. Probabilities of very rapid progress will fluctuate between 76% and 85%.

The higher rates of progress occur when there is a surge in the strength of some/all factors present and/or when stronger convergences of the factors occur.

## Soaring Level of Progress

*Children's House Drawing Stages.*

Here is a tricky mix that has been shown to support deep engagement, peak performance, breakneck learning, and cascades of progress.

1. Learner has seen new live drawings in progress and experimented in response
2. Learner sees some immediate drawing progress
3. Learner "shoots" the personal sacred cow of "I lack artistic potential"
4. An adult who is well-liked by the child directly encourages the child's drawing
5. An adult who is a highly skilled artist and an emotionally significant partner of Learner fluently makes their own complex, distinctive 3D drawings right beside Learner
6. One artist makes a variety of different 3D drawings as the child watches

7. A second and third artist is on hand and also does what is described above for the first artist

8. The anxiety is low

9. The artists make no direct commands or instructions to the child, instead allowing the child to choose what aspects of the artists' creations are of interest

10. Two of the child's friends are present and also share in the above experiences and are relaxed and enthusiastic

Probabilities of very rapid progress with this rich dynamic mix of ten factors (five dyads) will fluctuate between 85% and 99%.

The house in the above figure marked as #8 in complexity score was actually produced by a four-year-old after a few hours of a *soaring* interaction we arranged in our lab with a skilled adult artist. The nature of their interaction together encapsulates the multiple factors described above for *soaring* conditions.

## Convergent Force Fields within Dynamic Mixes

Consider any situation in which a clear goal has been set, whether for an individual or a group. Using the dynamic tricky mix approach, we can calculate how much "force" toward the goal has been established by the convergent factors at any given moment.

The total convergent force field will be the net effect in a multiplicative fashion of all positive convergent factors minus the combined effect of negative/oppositional forces. In a few moments, we will return to the dynamics of 3D drawing—first considering the broad applicability of this force metaphor. Note that, in all cases, force is not just a simple motivational concept of desire or effort by individuals, but rather the result of all engagement, focus, attitudinal, motivational, strategic, experiential, contextual, and related factors dynamically converging and emerging in events.

### Charging Bison Example

Suppose a bison in Glacier National Park has chosen the goal of ramming a particularly obnoxious and loud tourist. Suppose further that this bison has initiated movement toward the (stationary) tourist and has reached a decent speed and an accurate trajectory.

Total convergent force when the 2,000 lb. bison hits the tourist will be determined by the formula: Force = Mass × Acceleration (the square of feet/sec.). So, if the bison is walking at 2 feet/sec., the force on the tourist will be a modest 249 lbs. At that low force, you as the tourist might even be able to create 250 lbs. of opposite/negative force and stop the bison right in its tracks!

With greater acceleration, as the convergent force rises, reaching 4, 6, and 10 feet/sec. respectively, force on the tourist rises to 995 lbs., 2,238 lbs., and then 6,216 lbs. For the moment, we will leave open to speculation what mixes of conditions would arouse a bison to move at any of these illustrative rates of acceleration.

Similarly, in any kind of potential breakthrough project or event, radically different levels of convergent force toward a goal may be created by changes in just one or two factors.

### Runner in Olympic One-Hundred-Meter Dash Finals

Let's imagine seven finalists who are all uninjured, fully trained, and drug free for this event. Further, assume that there is no more than a 0.10 second difference in each of their best race times for the year.

What mix of conditions would lead to one runner rising to the occasion and winning the dash and international fame? We will revisit this question soon. But perhaps make a few notes to yourself on what mix you think would be powerful.

Please make your notes here.

## Introduction of a Specific Force Heuristic for Dynamic Tricky Mixes

Prior accounts of "force Fields" or similar concepts for human learning and performance have not done justice to the full range of force levels. Here, we will examine how force outcomes vary when we examine, in each case, one to five dyads of converging factors, each with a strength level ranging from two to six. Total force toward positive performance is calculated in a multiplicative sum across all dyads.

(In line with Dynamic Systems and convergent processes, a single component of any dyad will not exceed a trivial strength level of one).

For this Olympic dash example, let's assume that the winner and the runner-up each have five powerful dyads in place but with just slight strength level differences on one dyad.

> Preparation Strength = 6
> Strength/Flexibility Strength = 6
> Racing Savvy Strength = 6
> Rest/Sleep/Nutrition Strength = 6
> *
> Dynamic Monitoring of Self/Others:
> Winner's Strength = 6
> Second Place's Strength = 5
> *
> Multiplicative Force Sum of Winner: $6 \times 6 \times 6 \times 6 \times 6 = 7{,}776$
> Multiplicative Force Sum of Runner-Up: $6 \times 6 \times 6 \times 6 \times 5 = 6{,}480$

Thus, a small strength difference in just one dyad, when it converges with other dynamic factors, gives the winner the edge they need—perhaps a finish only 0.04 seconds ahead of the runner-up.

## Child at Flowing Versus Soaring Levels of Progress Toward 3D Drawing

Recall that, for a child in the *soaring* level, the probability of very rapid progress with this rich dynamic mix of ten factors (five dyads) will fluctuate between 85% and 99%.

Now we will calculate force in a fashion similar to the runner in the Olympic dash. Although the factors and dyads are quite different, we have five dyads where strength is near or at top levels. In the *flowing* state described earlier, substantial progress in drawing skill will be close to the probability of 85% when we have this outcome:

Multiplicative Force Sum equals $6 \times 6 \times 6 \times 6 \times 5 = 6{,}480$.

In contrast, progress probability approaches near certainty (99%) when just one dyad strength jumps up one level from five to six: Multiplicative Force equals $6 \times 6 \times 6 \times 6 \times 6 = 7,776$.

This clearly is the qualitatively-superior level of progress we call *soaring*. In quantitative terms, he force field is clearly far greater than that shown at the "moving" mode level described next.

## Progress in Leadership Facilitation Which Results in Transformational Public Policy Effects

In public policy toward nature, global warming, inequality, access to health care, reduction of toxins, and other key social issues, the dynamic processes are the same as in the above discussions and examples. Naturally, though, many of the particular dynamic mix factors will be different.

Here we will draw upon and extend the work of Monica Sharma and her colleagues as well as insights from Ashoka projects worldwide to lay out enough information to suggest scenarios of effective transformational chains and their relative total forces toward change.

1. Complexity is faced
2. Tools already available are adapted
3. Leader(s) roots planning in values
4. Leader(s) roots action/execution in values
5. Citizens recognize their own power
6. New tools are welcomed and tested actively
7. Multiple perspectives are woven together, including appraisal of seemingly "crazy" ideas
8. Rigorous small-scale experiments are analyzed both for strong process and strong outcomes
9. Confidence builds based on positive results and an understanding of why plans are working
10. Courage to confront problems is high

Convergence of these ten factors, in five dyads, will lead to different outcomes depending on the strength levels of the five dynamically synergistic dyads.

When all dyads are at level three, total force toward project progress registers at two-hundred forty-three ($3 \times 3 \times 3 \times 3 \times 3$). Most likely, this would show dynamics of an *Improving* Mode.

But for the very same five dyads, total force reaches a *soaring* level if two dyads rise to level five strength and three dyads rise higher to level six strength. Exciting, on-target, highly engaging movement toward goals will occur. Total force reaches 4,500 ($5 \times 5 \times 6 \times 6 \times 6$).

## Leading Corporations or Foundations to *Soaring* Levels of Performance

"If you have the right people on the bus, they will be self-motivated. The real question then becomes: How do you manage in such a way as not to de-motivate people? And one of the single most de-motivating actions you can take is to hold out false hopes soon to be swept away by events.

"Yes, leadership is about vision. But leadership is equally about creating a climate where the truth is heard and the brutal facts confronted."

Jim Collins, in this quote from *Good to Great*, sums up decades of studying and advising the corporate world.

His view meshes strongly with the many emphases we have placed—for any domain of activity—on valid monitoring of what's really going on, then adjusting, experimenting, and continuing to follow through. In contrast, we have seen countless examples of years being wasted on a committed path which—when finally monitored—proves to have been a total waste of time and resources.

So, let's now place some of the components Collins stresses for great leadership—along with other factors we identify—to illustrate a set of converging factors for a *soaring* level of organizational progress.

1. Leaders model "thirst" for brutal facts
2. Team members share facts as well as ideas freely
3. When failure occurs, the typical response is to analyze and adjust in a positive way.

4. Individual team members find their own niche, fitting their spirit and skills, whilst accepting common goals

5. "Experimental" frames compare the current approach to alternative approaches that appear promising but that await actual monitoring

6. At least 5% of the organization's projects are aiming very high for the "impossible" outcomes (breakthroughs)

7. Playful and fun events strengthen social bonding and shared positive expectations while also lowering defensiveness

8. Positive snowballs emerge with richer and richer ideas, experiments, continuing successes, and quality of monitoring information

9. Leaders as well as team members move forward and maintain engagement, primarily because of intrinsic motivation, but also to reap high external rewards such as recognition, income, and social access

10. "Wild Card Teams" are often formed in which a project in domain X (e.g., marketing, fundraising, writing code) employs wild card individuals who have skills/backgrounds different from the usual expectations of that team (e.g., composer, graphic artist, blues singer)

Now, let's say that all of these converge positively and with excellent, concordant timing. Great—a *soaring* mode has been reached! High performance will be in place and sustained as long as the conditions and their multiplicative convergences persist.

In this chapter, we have invoked the metaphor of "quantum leaps." The successive levels from *stuck* to *soaring* are not just simple steps up on a ladder of activity and progress—each new level is a leap to transformed intensity and complexity of the dynamic processes at work. By keeping such transformations in awareness, we increase the likelihood that our own re-mixing efforts may be powerful enough to facilitate leaps to outstanding breakthroughs.

However, we will see later that powerful *negative convergences* may emerge under other circumstances. For such instances, the dynamic processes and force processes at work must be recognized and offset if positive trajectories are to be re-established.

## A Few Caveats

The examples in this chapter of quantitative force equations/calculations could be approached using other metrics. Other authors are hereby invited to explore those.

However, several aspects of the dynamic tricky mix "quantum leaps" presented here should not be overlooked.

First, the account here covers strong variations in total dynamic force which have not been addressed well by other motivational theories. Recognizing and adjusting strategies for breakthroughs on the basis of such wide variations is crucial.

Second, combined dynamic force towards an objective includes dynamic interactive processes which integrate multiple, traditionally "motivational" conditions (e.g., self-efficacy) with multiple cognitive and social conditions. Among the latter, depending on context and objective, are mindfulness, positive expectation, low anxiety, social scaffolding between cooperating partners, awareness and focus, positive emotional engagement, monitoring of progress, and activation of component cognitive processes. Included in many highly positive "flowing" or "soaring" process levels will be blends and hybrids of two or more emotional states and two or more motivational proclivities.

Third, "total force toward project/objective progress" is the label we have used here for such broad dynamic integration across many complex conditions. "Total momentum" might be considered as a possible alternative label. Either way, motivation does not sit in isolation like variations in the strength of wind on a sail, but rather is always viewed within its diverse set of conditions and elements. In terms of methods to apply, variations on how total momentum is achieved could be modeled through multidimensional scaling techniques.

## Chapter 12

# WHEN IS YOUR BIRTHDAY?

You or someone you know has a birthday in June. Someone else you know has a birthday in October. Can you imagine a world in which there were huge differences in your life prospects that are triggered by those differing birth months? You would no-doubt be pretty surprised if, on average, June-born humans lived only 1/8 as long as October-born humans.

Well, surprise!

Because that kind of way-out, surprising difference in life spans holds, not for humans, but for a species of butterfly. Monarch butterflies.

Like many other insect species, the usual life span for monarchs is pretty short. For 90% or so of monarchs, one month is the expected life span. This is true for all monarchs born in June, July, and August in the United States or Canada.

The kicker is that October-born monarchs are wildly different in multiple respects. Yes, they live for about eight months. And their life pattern is remarkable. They escape a cold death by leaving the northern climes soon after they are born. They head south, migrating as far as Mexico and Central America. Travel south may take eight hundred to two thousand miles as well as considerable stamina and accurate navigation.

Despite having the same genetic makeup as June-born monarchs, these October surprises interact with the daylight, temperature, sun, and star patterns of their environment. Their bodies remain strong, vital, and young looking, for seven to eight months. Accordingly, they survive demanding travel and the winter in warm climes. They also survive further demanding travel northward in the spring to find flourishing milkweed. Mating behavior is then triggered and the females lay their eggs upon just one kind of plant—the milkweed. The adult males and females who have completed these missions soon die. As the eggs they left behind turn to caterpillars, and then to chrysalis, and then to a new hatch of adult monarchs, this begins a summer series. Wherever they are born, the subsequent generations move further north until monarchs are widely spread out across the northern United States and southern Canada.

October arrives, however, and late-season monarch butterflies emerge from their cocoons. The dynamic mix of genes and environment now leads to all the surprising characteristics of October babies. The same dynamic processes and principles apply, but because the resultant dynamic tricky mix of genetic makeup and environmental complexities are so different that the unfolding direction and timing of travel, delay of mating, and wintering-over in Mexico or nearby all cascade out for the remarkable eight-month, long-life adventures of October monarchs. Chapter 25 later expands on dynamic processes around the lives of monarchs.

# Chapter 13

# WE MET ONCE TWENTY YEARS AGO, BUT LET ME SHOW YOU I REMEMBER YOUR SHOES, BUTTONS, FOOD ORDER & WORDS

In my case, I can assure you that I do not find it easy to recall fine details about most events from many years back. Moreover, I am betting that also holds true for you, dear reader, as well as all your friends and relatives. No big deal, really, in terms of impact on our lives—we don't need that kind of memory achievement in many circumstances.

What is a big deal is that a small percentage of adults are fantastic in their long-term memory of details. An even bigger deal is that it appears that their brains are not really better or faster at attending to events or other basic components of processing information. That presents a profound paradox or puzzle for most theories. But it fits beautifully with the many other aspects of dynamic mixes that we are exploring in this book.

"Superior autobiographical memory" folks, SAMs for short, typically do not use that memory as the core of what they do for a career. A clear example is one woman who acts for a living and loves to collect shoes. Videos have captured her moving down the rows of shoes in her closets and answering detailed questions about any chosen shoe—when and where she bought them, what she paid for them, the first time she wore the shoes and with which other clothes on that occasion and others. Her memory of those kinds of details of everyday life events far exceeds that of her husband or young son, often leading to astonishment and sometimes to conflicts. After all, if you disagree with someone with such an exceptional memory about what happened in the past, it is hard to win a dispute.

Similarly, people with SAMs pull off many other feats of memory unavailable to most of us. For example, when asked abruptly to recall details of what happened in the world on a random date, they rapidly review detailed happenings of that date.

Memory systems for SAMs folks or for the rest of us do not proceed automatically. What is common for all of us is the set of cognitive processes that need to cooperate dynamically for long-term recall of events. As in so many instances already examined in this book, a few new twists on the unfolding dynamics can make all the difference between no success, occasional success, or a high rate of success.

For myself, I really do not care about when I bought a shoe or when I wore it. But for a SAMs person who does care about the small details of her experiences, there can be a long cascade of dynamic mixes that support exceptional long-term memory. By paying more attention than most to details as they happen, encoding these details more richly in memory by revisiting, rehearsing, and sharing these details in the days, weeks, or months after they occur, a shoe bought on October 1, 2005, enters neural networks of the brain repeatedly and the strength and extent of the memory patterns keep increasing. Strong, highly accessible memory traces are thus built up along with an appropriate confidence and enjoyment in the successful retrieval and re-experiencing of events.

Fluent readers use these same dynamic processes to support the basics of accurate, flexible, rapid, and enjoyable reading. When we become an expert in the domains central to our work and hobbies, the same impressive cycles of greater stored knowledge, skills, anticipation, confidence, and enjoyment in the encounters are put in motion. We also use fresh events supported by related prior events and contexts that have been retrieved from our long-term memory storage.

Similarly, most of us become "expert" at an early age in comprehending and speaking one to three languages. My colleague Marnie Arkenberg and I have argued that this viewpoint on early language achievement as expertise should be seriously discussed if we are to fully grasp what happens dynamically as we acquire our childhood spoken and/or sign languages. The dynamics of expert language acquisition and life-long expert language performance are illustrating once again the prevalence of dynamic tricky mixes at work. There's more to come along these lines in the following chapters.

## Chapter 14

# LATE-TALKING CHILDREN: BREAKTHROUGHS AND SLAIN SACRED COWS

B y considering in this chapter research on children's language delay and language acquisition, we will be able to show that, in these areas, oversimplifications have proved sometimes dangerous to children's well-being. Further, these oversimplifications contributed to poor scientific accounts of the phenomena of language acquisition.

In my lab, and in collaborative efforts with multiple other universities, we have often studied children who are about six years of age but whose language levels are delayed by three years—they are talking like typical three-year-olds. The first simplified conclusion that guided most earlier attempts to help these children is that they need special simplistic, didactic input. In language therapy, they were usually given imitation drills that are designed to be repetitive and make obvious to these "slow learners" that sentences have grammatical structure. So, for example, a child might try to imitate twenty sentences in a row that all call attention to *-ed* and the past tense. The dogs *played* with the bones. The girls *talked* on their phones. The boys *walked* across the grass. And so on.

A second simplifying assumption is that children who have been slow in learning language need their input to be only very slightly challenging. Therefore, the above imitation drill procedures focus only on structures that the child has already begun to use. Past tense examples in a treatment drill would be considered appropriate for any child who—in ordinary conversations—used *played, opened,* etc. but only 5% to 30% of the times where past tense would always be used by adults.

*Surprise!* When my colleague Stephen Camarata and I, along with talented graduate students, did a total re-mix of treatment procedures, plus appropriate experimental controls, we found that all of the above assumptions were wrong. Our research was supported by the National Institutes of Health. Moreover, the results of our series of causal experiments have been confirmed in new experiments by independent colleagues all over the globe.

One key to our research was that we knew ordinary conversations vary tremendously. So we borrowed clues from the richest and most challenging examples that we found parents using. We innovated by making rigorous procedures with three different kinds of experimental controls which looked promising for rich conversations. We theorized from Dynamic Systems thinking that re-mixed conversations would trigger rapid syntax/grammar growth by providing high challenges, high supportive conditions, and deep engagement by a child.

And that's exactly what we found! Children with poor histories of language progress now learned rapidly. These children were not learning from their ordinary environments, and even though they were two or three years behind, they *Soared* in language levels because of the new intervention. We labeled this new approach "conversational recasting."

More specifically, here are the conditions that mixed together for effective language treatment that was planned and executed within the dynamic tricky mix model.

1. Each child received intervention laser-focused on the language forms/structures they were lacking—these are child-specific targets
2. The clinician waited for the child to say something they loved to talk about

3. The clinician followed the child's utterance/comment with a recast, keeping the main topic/meaning but building in a child-specific target
4. The clinician/child dyad had fun together
5. Both experienced frequent positive emotion
6. The child showed deep engagement in the conversation
7. The clinician from the beginning of treatment expected good language progress despite the child's poor history of learning
8. The child began to learn in early sessions, and on that basis changed their own expectations for later sessions, expecting to learn well
9. Snowballs emerged where the longer the sessions went on the better the conditions—more and more positive emotion, attentional focus, fun, relaxation, and learning
10. No treatment time was spent on tiny challenges the child would learn without treatment—as clearly shown by control tracking where children not receiving treatment mastered these tiny challenges on their own within four months

In one sense, the review of these studies has been a sort of case study of over-simplistic thinking concerning what procedures can cause children with severe language delay—who were fundamentally *stuck*—to make significant progress in their sentence structures. The children's gains are important in their real-life, ordinary contexts of school, home, and play with peers. Children who had been behind by two to three years could not keep up with peers and were making no further progress, but eventually moved toward a state of communicative equality.

We shall see in subsequent chapters that strikingly similar general dynamic tricky mixing processes are at work in many aspects of life, both for adults and children.

Chapter 15

# THE SURPRISING DYNAMICS OF LONG-NECKED GIRAFFES AND FISH WITH NO RIBS

## How Do Giraffes Get Their Long Necks?

One answer to this question involves evolutionary time. Scientists agree that over millions of years, giraffes became different in their appearance and behavior from other African hooved animals such as antelope and zebras. But why? Theorists give various answers ranging from the evolutionary advantages of reaching high leaves on trees, better long-distance views to detect predators, or better battling of males against each other. Future creative thinking and testing of hypotheses may settle these open questions.

Another answer to our question concerns what happens in embryonic development before birth and then in early life as a giraffe calf.

Let's work backwards on this. A fully grown giraffe may have a neck that is eight feet long. You already knew it was long, but still, isn't that amazing?

Even more amazing is that the basic dynamic processes for growing a neck in a developing embryo and in early life after birth are always the same. For all mammals, the genetic codes for building vertebrae in the back and neck lie quiet until they turn on at the appropriate point in embryonic development. For the short neck of a developing mouse, it doesn't take long before the neck is built and the relevant genes turn off. The giraffe embryo uses the same basic mammalian genes and builds a long neck by letting those genes be active for a longer time. Of course, further growth in the size of the vertebrae occurs during the months between birth and adulthood.

Remarkably, though, the same fundamental dynamic processes for neck growth end up producing necks *in Giraffa camelopardalis* that are roughly two hundred times the length of the neck of a common deer mouse, *Peromyscus maniculatus*.

We might compare outcomes like these to the old goals of the alchemists to turn a less valuable metal like lead into gold. In this instance, we have the ordinary "lead" genes, and these yield, through the "alchemy" of dynamic gene interactions, 5,000 species of mammals with all their marvelous variations and beauty. This is a new, objective sort of alchemy creating breakthrough transformations that lie at the heart of animal biodiversity.

Keep in mind for later comparisons the magnitude of effects that can occur from unfolding dynamic processes. Recall that, for monarch butterflies, October butterflies live eight times longer than mid-summer butterflies. When we look at neck growth in animals, we see the huge variations in neck size unfolding from the identical underlying processes. All it takes is a tricky catalyst called "timing control" governed by regulatory genes that stop neck-building at just the appropriate point for each species.

## How do Leopards Get Their Spots?

Similar examples occur in explanations of why the coats of animals look so different. To a common observer, it would be easy to assume that fundamentally different "architectural designs" somewhere in the genes would lie behind spotted leopards, striped tigers or zebras, horses with pinto coat patterns versus solid bays, vivid white-and-black cattle versus solid blacks, and so on. In other words, a simplistic (but wrong) assumption is that somewhere in the genes a pattern of what an individual animal should look like has been stored and then is activated to make an animal's skin pattern.

Geneticist Sean Carroll hits hard against any simplistic account for different forms of animals and different behavioral traits:

"There is no need to invoke single dramatic mutations as causes of great leaps in form and function or as explanation for the origins of human traits. Nor is there any scientific foundation for doing so."

Instead, for mammals of all kinds, the same genetic "coat painting" mechanism kicks in during embryological development. Then, the unfolding dynamic processes require just a tweak (regulation) from the genes turning on and from how the pigmentation flows on the "environment" of the growing animal's skin in the embryo. Thus, in a litter from parent dogs of the same coat color, we may see the emergence of only golden Labrador retrievers, only black ones, only chocolate ones, or any mix of these. The same goes for

Part Zebra ... Part Tiger? Nope. All Tiger—Just One with a Small Dynamic Tweak in Embryo for Coat Design.

all those colorful coat variations on cats, horses, cows, and goats. In effect, the equivalent of a few "pigment spray guns" turn on and turn off in sequence to create areas of "paint" flow that dynamically influence each other. For example, these tricky mix dynamics will "paint" a spotted clouded leopard with its distinct coat. Or, in the instance pictured below, create the seldom-seen albino version of a tiger with black stripes on a white background.

## Surprise: A Fish with Missing Ribs

All species of pufferfish are able to rapidly fill their bellies with air or water, expanding like balloons to twice or even three times their normal size! In Dynamic Systems terms, their evolutionary history has somehow selected for this trait. And, as is always true in evolution, the emergence of this special trait was accompanied by a chorus of additional traits—some essential to puffing up, and others mostly independent of puffing. Regarding independent traits, pufferfish are the second most poisonous vertebrate, right behind poison dart frogs, thanks to concentrations of a toxic chemical, tetrodotoxin, in their skin and liver.

Now, for a small package of traits that work together in pufferfish, consider this species: dog-faced puffer, *Arothron nigropunctatus*. This guy grows to about one foot in length. A crucial feature that makes puffing up especially efficient is that it has no lower rib bones or pectoral fins, allowing it to rapidly swallow large amounts of water or air to fend off predators. A further, related part of its defenses are sharp spines under the fish's skin which can pierce predators when the body is inflated.

We can easily see that such pufferfish are quite exceptional in the fish world. That exceptionality is something we want to be sure to connect to multiple strategies for helping to create breakthroughs. Any time, in any domain or field, where we can identify remarkable exceptions to the norm, exceptions to typical cases, we should be alert. What are similar processes in other fields or events to the exception we have just found? Here, a fish is not fish-like in its number of ribs, and that lack facilitates unique behaviors.

So . . . could we be good tricky mix experimenters, always keeping dynamic complexities in mind, and set up novel sets of conditions which create never-before-seen exceptions in a totally different field—e.g., in pollution control or tools for fighting pandemic viruses?

# Chapter 16

# ALL CALM IN THE DENTIST'S CHAIR

n television ads, dentists assure us that if we come visit them we will have a calm, peaceful, positive experience; caring, sensitive, feel-good dentists.

Somehow, that doesn't square with most people's experiences in dentists' chairs. For many years, the adjectives I would apply to my dental adventures are "uncomfortable," "painful," and "yucky." The pain included the jabbing needle of the local anesthetic as well as further pain before I was completely reacting to the injection. After getting a cavity filled, I would walk out with a distorted, mostly numb lip that led me to slobber my drink down my chin and onto my shirt unless I took extreme care.

How could things go better? A re-mix toward the positive could not leave out what the dentist needed to do with his drill, lobster pick (just kidding), and so on. Nevertheless, I discovered a breakthrough path.

While teaching a special graduate seminar in imagery, we ran across cases where it was claimed that minor surgery and dental work had been approached through active imagery. The imagery was reported to shift the patients' attention so strongly that they experienced no pain, remaining calm without the assistance of any painkillers.

Since I had a tooth that needed attention, I decided that I would try imagery and report back to my students. I had already been practicing imagery for the meditative effects it can promote. Through such imagery I had reached states of high relaxation and the flow of many beautiful, vivid scenes in my head.

So . . . when I was in the dentist's chair and the dental assistant asked if I was ready for my needle, I said "No, thank you. I will just use imagery." She and the dentist were skeptical indeed. They consented to proceed but had the hypodermic filled and ready and told me that "at any point" if I needed the needle to just speak up. As the drill began to grind away, the vibrations and sounds fed right into my imagery. One scene in my head was the feeling of one of my favorite sailboats with its vibrations and movements as it rose and fell and moved through wave after wave of ocean saltwater. I could smell the salt and feel the spray as well as the joy of guiding the sailboat through puffs of wind and the peaks and troughs of the waves. I breathed deeply and calmly and flowed along with the sensory input of my imagined scenes. Dynamic, tricky, complex re-mixing! The objective sensory input from my mouth of what the dentist was doing became part of the rhythm and pattern of the sailing scene, and none of the dentist-produced input was felt as pain. Beautiful!

Upon leaving the dentist's office that first day, I was happy to find myself bouncing along with a refreshed feeling much like a good session of meditation anywhere. No numb lips, no slobbering, no negative feelings at all. I reported all this to my students, shared it with friends, and vowed to continue use of imagery if future dental visits came up. And when such visits did arise, I again went deep into imagery, avoided any needles, and found again that I felt glowing and positive and joyous feelings for hours after my imagery/meditation in a dentist's chair. Marvelous re-mixing of what occurs when visiting a dentist!

## Buddhist and Other Writings on How to Re-Mix Our Responses to Strong Stimuli

Of course, if we look back at ancient writings, we find evidence that these same dynamic processes for changing our interpretation of pain and other stimuli were actively pursued. Buddhist writings quote the Gautama Buddha on how we may exacerbate our own pain and suffering as well as how we may instead minimize the impact of pain on our lives. Here is that parable.

## The Arrow

"The well-instructed disciple of the noble ones, when touched with a feeling of pain, does not sorrow, grieve, or lament; does not beat his breast or become distraught. He feels one pain: physical, but not mental. Just as if they were to shoot a man with an arrow and, right afterward, did not shoot him with another one, so that he would feel the pain of only one arrow. In the same way, when touched with a feeling of pain, the well-instructed disciple of the noble ones does not sorrow, grieve, or lament; does not beat his breast or become distraught. He feels one pain: physical, but not mental.

He discerns, as it actually is present, the origination, passing away, allure, drawback, and escape from that feeling. As he discerns the origination, passing away, allure, drawback, and escape from that feeling, no ignorance-obsession with regard to that feeling of neither pleasure nor pain obsesses him." (Gautama Buddha)

And this, in turn, leads us to a modern experimental demonstration of relevance.

## Modern Research on the Power of Meditation

What if we look at true experts in meditation (which I have never claimed to be)? Buddhist monks with twenty or more years of practice and with practice in teaching meditative techniques to others have recently been studied using brain imaging techniques by Richard Davidson and his colleagues. The basic idea is to see with fMRI imaging techniques whether the "startle" and "pain" brain areas that respond to very loud, sudden sounds in most people would fail to activate in the monks. The experimental setup guaranteed that the objective loudness and timing of the sounds was the same for the monks as for the college students (non-meditators) being tested, as was the context of lying in a cramped position inside the fMRI scanner. The expert Buddhist monks began their meditation and signaled when they were deep into the meditation. At unpredictable intervals, the loud "startle" sounds were played, but the monks were dynamically re-mixing how their brains responded to the incoming sounds. They simply noted the sounds but felt no startle or discomfort, and their neural networks in their brains similarly showed calm, low activation patterns.

In terms of processes, this is the same kind of re-mixing that I did in the dentist's chair in my imagery meditations. Pain, surprise, and discomfort are not states guaranteed by a stimulus—they are always psychological states that depend upon the dynamic patterns that mix ongoing thought/feeling patterns with the incoming sensory inputs. In modern psychological and neurological theorizing, my experienced states are always "embodied." They are never independent of what is already ongoing in my body and brain when new information, new stimuli, or new opportunities arrive.

## Invitation to the Reader: Try This Experiment to Achieve a Tricky Re-Mix of Your Responses to Strong, Persistent Stimuli

Consider the possibility that, in some respects, you have been "shooting" yourself needlessly with arrow after arrow. This has definitely happened to me, and to most people at times.

Let's imagine that there was a time in your past when you let a certain opportunity pass by so that you could pursue other possibilities. Then, later, you recall the situation and "shoot" yourself with a painful arrow of regret. This pain might easily pass by. But if you repeatedly bring this regret to mind, the pain may increase over time and you may find yourself thinking way too often about it, causing considerable distraction and stress.

Please try out the experimental approach of a tricky mixer through a thought experiment. See if you can lay out six active strategies/steps you could easily test out as possible means for reducing or eliminating the continuing negative influence of this past event on your current life.

## Accessing Images from Nature

In other chapters, the multiple ways that nature experiences can help create new conditions that feed into positive mixes and breakthroughs will be discussed.

In the context of the current chapter, consider that deep immersion in the beauty and flow of nature may sometimes help us remove an "arrow" of irritation and/or preoccupation. Maybe you have already discovered a special place you revisit from time to time through imagery to seek peace and inspiration? If so, consider expanding on such imagery strategies, expanding and deepening your mix of meditative resources. And if such imagery exercises would be new to you, explore and experiment until you identify nature scenes that work well for you.

Now, as a break from reading, sit quietly and examine and absorb the image below. Let your attention and your spirit move into the scene and imagine yourself becoming, in turn, the flying hawk, the meadow grass . . .

*Red-Shouldered Hawk Hunting and Flying Very Low.*

Or, let yourself be surprised by the next image, where the water of a pond has been transformed by light and ripples into what seems an entirely new and magical substance.

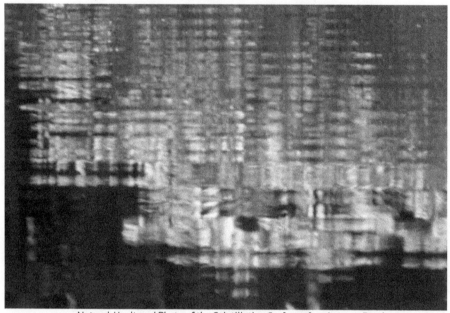

*Natural, Unaltered Photo of the Scintillating Surface of an Autumn Pond.*

## Chapter 17

# SEEK COMPLEX CHANGES IN SELF RATHER THAN SIMPLE KNOWLEDGE OR SKILL INCREASES

"It is not necessary to do extraordinary things in order to achieve extraordinary results."

—Warren Buffet

"To achieve extraordinary results, it is necessary to bring a mental model—although imperfect—of relevant complex conditions in dynamic interplay, and to make action proposals respecting those dynamics."

—Keith Nelson

"When you don't know what you are missing, it is best to explore widely and to employ a 'banquet' approach."

—Keith Nelson

In educational and clinical practice with children or adults, professionals often proceed with a preset plan for the sequence of steps forward in knowledge/ skills that seem appropriate. The desired sequence forward seems to be obvious, simple and based upon attentive planning. Instead, we have argued and illustrated already in many ways that moving forward is far from simple. Progress depends not just upon encounters with challenges, but on the real-time, concurrent, dynamic convergences of many social-emotional, motivational, and attitudinal conditions.

In this chapter, we explore new cases which demonstrate the importance of leaving space and flexibility in teaching encounters, creative endeavors, and corporate moves so that learners/clients are engaged and transformed. In seeking such complex changes, the talented clinician, teacher, coach, or other "change artist" prepares the learner/client to be more effective in the future while also advancing in desired skills and knowledge. Likewise, as an individual, we may consciously set out to be a "learner" of new ways, new mixes, new exploratory "banquets" that lead to personal growth.

Fundamental to all the cases we will now consider is the inescapable impact on future dynamic changes in a person's self-concept and expectations. Our re-mixed self will shift the dynamics of how new situations unfold. One factor to consider is the very "stances" that we take in our own life and other lives we may seek to enrich. Here are excerpts from a Billy Collins' poem that touch on these issues.

### My Life

Sometimes I see it as a straight line
drawn with a pencil and a ruler
transecting the circle of the world

. . . . . . . . . . . . . . . . . . . . . . . . . . . . .

Like yours, it could be anything,
a nest with one egg,
a hallway that leads to a thousand rooms—
whatever happens to float into view
when I close my eyes

. . . . . . . . . . . . . . . . . . . . . . . . . . . . .

I am a lake, my poem is an empty boat,
and my life is the breeze that blows
through the whole scene

. . . . . . . . . . . . . . . . . . . . . . . . . . . . .

stirring everything it touches—
the surface of the water, the limp sail,
even the heavy, leafy trees along the shore.
(Billy Collins, "My Life")

Changes in self, yes! But which aspects? And by what processes and by what triggers?

Whether or not we are in therapy or are part of corporate teams with focus on creativity and change, everyone can learn something from long-time experts in these areas. What we will soon see is that the way forward to significant change can be achieved with startlingly different dynamic mixes.

## Psychotherapy Guru James Bugental

I had the good fortune to meet and learn from James Bugental. He was already famous and influential as a founder and practitioner of existential humanistic psychotherapy.

First, we will look at some of his central ideas, and then at examples of changes induced in clients—including me.

From our dynamic tricky mix perspective, Bugental is a remarkable innovator who richly appreciated and dealt with the many complex components of personal change. His approach ruled out the simplistic goal of the therapist "curing the client" of a particular malady. Instead, he set out to create complex new mixes of conditions which lead the client to revitalize their own lives. Within this framework, he was able to help clients through his own direct psychotherapy sessions as well as through the training of other therapists who also implemented the existential humanistic approach.

Bugental's belief in the healing power of inner awareness requires a therapeutic approach that heightens the client's subjective awareness in the moment. To do this, Bugental focuses almost exclusively on his client's intrapersonal processes and inner stream of awareness. He emphasizes that the skill of actively self-monitoring the inner consciousness which is gained by the client through therapy is likely to enrich life long afterwards.

For one woman who was upset that each of her relationships ended in a sudden and puzzling fashion, Bugental drew her awareness to what was happening right there in the therapy room with him—namely that she wasn't fully present and not highly engaged with their exchanges. Instead, she was letting her attention and feelings drift to past situations and scenarios in her head of what she could have done differently. By bringing her back to the present moment, as each therapy session unfolded, Bugental equipped her to genuinely experience and respond to what a partner says and does. This then led to her ability to form new and richer relationships outside of therapy.

## Experiencing Parts of Self Exercise

Bugental observes that our selves are complex and dynamically evolving structures. Further, we sometimes become disconnected even to those parts of self we consider our most central and important. Paradoxically, after many years of acting in line with these central parts, they may become too familiar and taken for granted. We fail to keep them in focus and alive, and may not even realize that this is taking place. Now I will share with you the essence of a workshop experience organized by Bugental to freshen our awareness and vitality. As the reader, I will ask you to participate and to record your own thoughts and feelings. Please find a partner who will agree to sit with you and enter into a shared exercise on what matters most to you as a person.

1. List here or on seven slips of paper seven parts/aspects of yourself that you consider most central to who you are. Choose parts that you know are central even if other people may be unaware.

2.  Share this list with your partner and vice versa, then discuss. Explore why each of your choices is an essential and enduring part of your makeup.

3.  Now consider the relative importance of your choices. Even though it may be difficult, place a #1 beside the most important, a #2 beside the next most important, and so on until all are ranked.

4.  This step is one where you imagine a future where you must for some reason experience the loss of each part, giving them up one at a time. First, imagine yourself without the least important of these essentials, the one you ranked #7. Take a few minutes to feel what your life would be without that self-component. Now repeat these steps in the order of your rankings, from rank #6 to the most precious component you ranked in first place.

5.  Then, go further and take a few minutes to feel what your life would be without all seven of these crucial components.

6.  Here, at this step, you have the chance to feel yourself come newly alive. Losing those seven parts of your complex self was only through imagination. If playing the violin has been a key part of you, when you put that down as a loss in fantasy, it no doubt felt sad and/or painful. Now, though, with new awareness, you will come back to reality and reappraise yourself.

*Caution!* Be prepared for some possibly intense emotions as you proceed. And with your partner agree to take whatever time it takes to process and relate to these emotions.

My personal experience with this situation proved deeply emotional and enlightening. James Bugental himself was leading a group of about twelve of us in the above steps. It is fair to say that everyone in that group was surprised at how such a brief exercise in imagination would lead to considerable tears, sadness, and increased awareness of one's own self-concept and life path.

Take back the component you ranked lowest, as #7 in importance. Take a few minutes to feel what your life can be in the future if you fully value and engage that self-component. Will it be an even more vivid and precious part of your life than it has been in the past?

Then let the process of recovering all the treasured parts of yourself continue. Take a few minutes to feel what your life would be when you again have self-component #6. Now repeat these steps in order for all your remaining rankings, one at a time, from rank #5 to the absolutely most precious component you ranked in first place. After each step, absorb and reflect on your feelings. See if there is a new sense of appreciation for all that you have going for you, for the richness of your life and experience.

## Why "Absurd" Team Composition Works So Brilliantly in Large Organizations

I once proposed to a college dean that he adopt a radical requirement for new hires in each of his twelve departments. This new plan would require that 20% of all new personnel be *unqualified*. That is, these new personnel would not have the experience, degrees, and networks usually considered necessary to be considered for the job. For example, someone would be hired in a chemical engineering department who brought something fresh to that department; someone who would teach, solve problems, and form collaborations in new ways. Perhaps an evolutionary biologist or a dance choreographer?

In a moment, we will see that a few corporate teams have fully implemented the kind of "absurdist" diversity I was inviting the dean to consider.

First, though, I will recount from my own research collaborations on helping deaf children begin to overcome their tremendously inadequate reading and writing levels. The breakthrough we aimed for was the creation of innovative tricky mixes that included many brief but rigorous tests, challenging lessons, and teacher-child interactions based upon high positive emotionality coupled with exceptional levels of sign language fluency for the teachers. We knew that this form of rich dynamic mix had been missing from the classrooms of deaf children for many years. And my colleague Philip Prinz and I were excited to create such mixes and see if we could overcome the very low, *stuck* levels of literacy of the ten-to-eleven-year-olds our project was funded to help.

However, we hit a snag. In recruiting teachers, we could locate no persons with the teaching experience and crucial sign language fluency we needed! So, we instead shifted gears and recruited excellent signers with no teaching experience. Absurd? Nope. Because with outgoing, personable signers we were able to quickly train them in all the teaching strategies of relevance. Our mix was further enhanced by software we created which incorporated—for the first time in the field of deaf education—reading and writing lessons with sign language appearing on the computers along with graphics and text. So, we put our innovative tricky mix in place for literacy lessons the likes of which the children had never encountered. And it worked well even in the first few weeks of instruction! The deaf children showed that they were capable of reading even though they all had been *stuck* at primitive levels of reading for two-to-five years.

For persuasive results and strategies, convergence of positive outcomes across research teams, different set of students, and different counties is paramount. For our tricky mix of innovative software and teacher strategies, such convergence stretches from the USA to Spain, Belgium, Norway, and Sweden. Among notable collaborators in Sweden, we have been fortunate to have the central involvement of Mikael Heimann, Tomas Tjus, and Mats Lundalv. Tobii corporation now carries a commercial version of this software for facilitating children's literacy and language entitled AnImega.

## "Design Thinking" and "Radical" Teaching/Coaching Approaches that Achieve Complex Changes in Student, Client, or Co-Worker Self-Concepts

Re-mixing complex conditions with close attention to the dynamics of intrapersonal and interpersonal conditions is at the heart of shifts toward more effective, and even transformational, changes during teaching, therapy, or life coaching.

Under consideration next are how different teachers and corporate innovators bring into play new mixes of awareness and experimentation that are strikingly similar to the therapeutic moves and strategies we have seen with Bugental.

David Kelley is both a teacher of design thinking at Stanford University and a corporate coach (see below). One hallmark of his approach is to shake up, to re-mix, the process of observing what is already taking place in any context. When team members are widely diverse in their training and backgrounds, that by itself will ensure that not everybody sees the same events the same way. For example, an engineer, a psychologist, a drummer, and a biologist working together may be productive despite conventional tendencies to rely instead on teams that are homogeneous. Observation techniques are steered toward active exploration of varying viewpoints as well, with small experiments to see what happens when something new is introduced. And rather than just documenting what folks are doing, the observer always looks for reasons behind the actions. Whenever possible, team members should seek to find what happens under unusual circumstances. Altogether, the "design thinking" approach is likely to generate innovative hypotheses on how a re-mix of standard procedures for a bank, a school, a hospital, a corporate marketing team, or other contexts could lead to more experienced vitality of participants, fewer negative side effects, better performance, and, yes, breakthroughs.

## Christopher Uhl & Dana Stuchul: Authors of Teaching as if Life Matters

Like Bugental, these authors know that profound changes for teachers along with their students require fundamental and radical shifts in awareness. To do a successful re-mix of learning conditions requires recognizing and moving beyond the too-common toxic mixes that pass for school instruction.

"The sad truth is that school, with few exceptions, is enacted in such a way that it mostly distracts young people from what is compelling, immediate, motivating, and engaging—thereby undermining self-discovery in any significant sense. Indeed, the idea that young people should have a significant say in their lives—that a ten-year-old should have the freedom to choose whether to go to school or not, as well as what to learn and how—is regarded in many quarters as terribly naïve. But the truth is that when we accord deep respect to young people, when we believe unconditionally in their innate intelligence and goodness, and give them a significant say in their lives, what happens is that they shine, growing daily in brilliance, self-respect, creativity, and power."

## Jack Matson, Founder of the Leonard Center for Entrepreneurship at Penn State University and Author of Innovate or Die

"After teaching eleven courses and completing a number of personal innovative projects, the process is becoming clear to me. I am not a professional innovator; I am a teacher, but innovation is my way of life. My reasons for wanting to be innovative are: mental and psychological health, the challenge of competition, the satisfaction of accomplishment, and the pleasure of knowing I am a live, vibrant, contributing human being. Reaching some of these goals, using the innovative process, has taught me that living a creative, innovative life is possible, realizable, and exciting . . . The use of the creative powers within you can lead to a most beneficial and wonderful phenomenon, that of inventing your own future. Instead of being at the mercy of outside forces, you can turn it around and mold those forces yourself."

## Corporate Coach for Innovation, David Kelley

Some of the most intriguing examples of wildly diverse, nearly "absurd" combinations of varied talents on the same team come from the work of David Kelley and his colleagues. They have done myriad projects from the launchpads of Ideo Corporation and Stanford University. They stress a series of interrelated strategies that we will recast in terms of paths to creating highly original dynamic tricky mixes.

Consider a project where a large airport or a large hospital seeks to modernize in ways that enhance their central services and also improve the pleasantness of the experience for customers/patients as well as employees. Kelley and his colleagues would choose an apparently "absurd" mix of team members to guarantee a diversity of idea generation and problem-solving approaches. Opera singers, anthropologists, drummers, and chefs would all be of value in this framework. They would join teams that also included some persons with prior direct work on airport or hospital projects. The resultant mix of team members led successfully, in project after project, to a bubbling mix of ideas generated, experiments for testing ideas out, and revisions of plans based upon careful and valid evaluations of what was working well and what was not.

We can speculate about how, in the future, creative teams might lead to additional breakthroughs. Consider what would happen to announcements over airport speakers urging particular travelers to hurry to their departure gate for an imminent flight. Those announcements are a simplistic attempt to prevent missed flights based upon false assumptions.

A creative tricky mix innovative design team would be sure to include video records of what staff do at departure gates, cellphone records of when and where travelers check updates about flights on their phones, interviews with travelers, a diverse set of team backgrounds, and, of course, data on how often flights are missed once travelers are in the airport. Here is a speculation on a shocking finding that is likely to emerge: oral public announcements made in typical dispersed messages to all areas and all travelers in the airport *decrease* the chance that travelers will make their connection by hurrying to a gate within the last ten minutes before departure.

So, dear reader, before reading on, reflect on why this conclusion is likely once careful observations are made.

We know from other research that reminders about upcoming events/obligations often work. The key question to ask is, when—that is, under what mix of conditions—are they effective? For upcoming dental or physician appointments, reminders are most likely to work if they are salient and also timed within two to seven days before the event.

And now, back to the airport. Modern travelers rely on a bevy of digital tools at their fingertips. Among these tools are messages from airlines and from fellow travelers in terms of emails and texts and voicemails. These are likely to be salient and timely. In contrast, the repeated and widely broadcasted announcements in the airport about irrelevant flights will lead the traveler to tune out and ignore them. Moreover, the blare of these grating messages over and over will tend to make the traveler irritable and less able to concentrate fully on what they need to do. So, paradoxically, the urgent alarms against missing a flight will lead the weary traveler away from effective use of the information they already have digitally and, therefore, will reduce the probability of on-time boarding.

## Summary of Self-Change Theories, Approaches, and Cases

We have seen in this chapter many contrasting human contexts and considerable contrasts in theoretical emphasis. Yet, emerging from these are a number of important common elements of significant long-term progress:

- Awareness shifts along with an increased interest in finding further awareness gains in the future
- Advances in communication that then feed into higher successes in both social and cognitive endeavors
- A new openness to actively seeking out new sources of information and strategies, trying out new experiments and projects, and more active monitoring of what is working and what is not working
- Attention to how one's biases, anxieties, illusions, or defenses may be blocking awareness and flexibility and a desire to reduce those negative influences
- Increases in a sense of self-efficacy—with all the components therein, including high confidence, positive expectations, high self-worth, and the realistic knowledge that, in the past, important obstacles were overcome and active explorations paid off richly

Valuable ways of exploring these processes can be found in the online *Coursera* course created by Jack Matson, John Bellanti, and their colleagues called Creativity, Innovation, and Change.

## Chapter 18

# OCEAN SAILING

The Magellan Axiom from Keith Nelson states: "When you enter with openness any context where you will often be in 'unfamiliar waters,' the dynamics will be highly favorable for genuine discoveries."

If you ever have a chance, try sailing in a small sailing vessel across stretches of open ocean. Now, you may be saying to yourself, "Isn't this a digression from thinking about dynamic breakthroughs?" But, as this chapter unfolds, trust that you will see yet more depth and detail on how similarities in dynamic patterns and advances arise across wildly different domains of experience and goal-sets.

When you go ocean sailing many miles offshore, you are likely to encounter some remarkable variations in how dynamic mixes and re-mixes occur. Your own perceptions and reactions will be re-mixed with unpredictable patterns of waves, currents, clouds, wind, storms, and wildlife.

I have had many ocean experiences and will share with you just a few that involve intriguing mix processes.

## Sailing Dramas in the Bermuda Triangle

A few years back, I joined a close friend and three other sailors to move a thirty-nine-foot sailboat from Tortola in the Caribbean back to Virginia. Our projected route should take us through the Bermuda Triangle for about ten days of

Atlantic Ocean sailing. We provisioned and tuned the boat and set off with high energy and enthusiasm.

Winds were favorable, and we moved along in fine fashion for the first forty-eight hours. This was followed by a day where the wind was light, fluky, and unpredictable. We had expected this and patiently accommodated to our slow progress. But then the wind fell to almost nothing. And then to absolute zero. In open ocean, we had not anticipated this state of affairs. We scanned the horizon in every direction, hoping for a hint of wind rippling the surface somewhere. No luck. So we fired up the engine, knowing we could make at least a little progress and perhaps move into some patches of wind. After twenty minutes we saw no trace of wind and decided to save our scarce fuel for any future, more critical, events.

There we sat in total quiet and solitude. For the moment, there was no sign of life in any direction. Flat, glass-like water. Our gallant sailboat floating listlessly like a child's bathtub toy. The initial frustration gradually gave way to laughter and surrender. Okay, Okay, we will just wait here and make the best of the situation. Soon we were swimming happily around our mid-ocean, Bermuda Triangle swimming dock. Ahh! Supremely relaxing.

You might think, "These guys are really in *stuck* position!" If our only goal was to make steady progress in this open-ocean crossing from the Caribbean to the Virginia mainland, then, sure, we were completely stalled. Fortunately, exploring whatever the ocean offers was also among our goals. So, without regret, we *pivoted* toward working openly with the opportunities our becalming offered.

Having a stable boat on still water, no rocking or bouncing whatsoever, set up a great opportunity to do some special navigation. We had a sextant onboard, together with all the relevant charts for interpreting celestial bodies. Given an accurate time of day, the sextant reading, and the charts, we would be able to determine our position. First target: the sun! (Safe to shoot through the sextant without losing an eye.) The initial data we got indicated that our position was Paris, France, rather than somewhere in the Bermuda Triangle off the Atlantic coast of the United States! Oops. This was not too surprising given that we had never even held a sextant before.

So, we refined our technique to try for a more accurate measurement. After some practice, two of us were getting consistent readings of angles off the sextant.

Because I had only heard of shooting star positions at night with sextants, I was amazed that the sun shots worked out well. Then, further fun came as night fell and stars were available.

Really, our perception of stars entered new sensory territory—a mind-boggling myriad of stars appeared in wonderful clarity compared to being on land close to unending electrical lighting of all kinds. Getting star positions was fairly easy now that we were practiced, and we just had to take care that we were identifying the proper name for the star we shot so that our charting would be sensible. Now we felt kinship with and respect for the ancient mariners who discovered continents, islands, and reefs with a tiny tool kit of navigation procedures. At the same time, we knew we were in perfect visibility and completely calm conditions. So it is remarkable that, despite clouds and rain and wildly rocking ships, enough data for navigation could be obtained to support safe passage in the 1600s.

The next morning, in soft dawn lighting, I was the first one up and had an entirely new view of this planet. In every direction was glimmering silver, calm water. When I moved my attention gradually out to the horizon, the earth absolutely *curved* down as if I were in the middle of an upside-down bowl. This induced a sense that I was only a speck within a giant, ordered universe. Yet, at the same instant, I was a speck full of powerful energy, with a new connection to the universe and its mysteries. There was so strong a clarity in the all-surrounding ocean water that it seemed as if the ocean had become pure light and that sailboat and sailors were lightly suspended on it. Total silence enhanced this illusion. Such grace, beauty, and energy in this event! What a dramatic re-mixing of my perceptions of water, light, and the connectedness of humans to the natural world!

"Take each day on its own terms."

"When the universe gives you lemons, make lemonade."

"Live in the present, in the now."

Similar advice can be found in a myriad of cultures and in spiritual, psychological, philosophical, and novelistic genres. Open ocean sailing prompts actually accepting this advice. By acceptance, the details of each day become more vivid, interesting, and satisfying. Once in the open ocean, there is no quick retreat to some other place that might cross your mind . Upon surrendering to the inevitable variability of circumstances that arise, a deep centering and peacefulness can be entered. In turn, such centering supports even more vivid attention to sensory encounters with wind, sky, water, and wildlife, as well as every aspect of the sailboat's movement and your own contributions to tuning the boat's trajectory.

With these thoughts in your mind, consider the events of our next day of sailing. Gradually a light wind picked up, then a little more, and then more. The joy of moving and feeling some breeze in our face after such completely calm, flat sea conditions brought us all to a mellow state. We held a simple easy sail for a while, with only the slightest physical effort needed to keep the sails and rudder in their proper positioning.

Conditions advanced even further toward the ideal. The wind grew strong, but not tricky to handle, and we could sail directly northeast toward our destination of Newport, the famous sailing port in Rhode Island. Moreover, we were on a "reach" sail setting that kept the sailboat nearly level and gave us steady, exciting power through the modest waves. The skies were clear and we zoomed along, high on these exciting, perfect conditions for hours.

From time to time, we had the further joy of seeing up close something we had only read about—flying fish. Something about our steady course and the only slight tilt, or "heel," of the boat made it relatively easy for the flying fish to rise up on our port side and sail right by our noses and disappear into the ocean on the right, starboard side. These fish were gliding rather than flapping to fly, but so what! They were smoothly airborne and as marvelous to watch in their own way as any hummingbird, swallow, or bluebird. Apparently, our sailboat on this trajectory was a perfect mix of conditions to stimulate these flying fish to launch themselves out of the water and across the boat.

The sun sank with a beautiful glow and the stars began to sparkle here and there in the sky. When full darkness had fallen, with no man-made light pollution to create interference, we moved along under a full three-hundred-sixty-degree dome consisting of stars of every magnitude. The brightest were not only vivid in the night sky, but also reflected in the water all around. The front of the boat, the bow, plowed along gaily through the star-strewn water and sent a steady river of water, a bow wave, moving from front to back along the port and starboard sides. Could this get better? Well, yes, remarkably it did get much better. We hit a stretch of ocean filled with tiny glowing, phosphorescent creatures. Our powerful bow waves now caused thousands of silver jewels to appear alongside our boat just a few yards from us. Our eyes were filled with a gorgeous river of stars when we looked up, and a gorgeous river of phosphorescent diamonds when we looked down. At some angles of view, both rivers of lights could be taken in together, both streaming toward us and disappearing behind us. Then, at times, the sparkling lights emerged into larger patterns—angels fluttering, dancing past. We sensed we were part of a larger angel as seen from the skies above, a large angel made up of our sailboat with one set of wings being our white sails and another set created by the two amazing streams of light created by our prow cutting through the water. The whole universe felt close, safe, and at the same time wildly dynamic and powerful. We flowed through this spectacle emotionally and visually, surrounded by moving air and water and specks of silver light, supremely content, alive, and in awe. What a remarkable mix! From time to time, we even remembered to breathe.

*Luminescent Nighttime View of Ocean.*

## Another Trip, Another Storm

From Tortola to Virginia, that sailing itinerary through the Bermuda Triangle was completed with the variations just described.

In contrast, one sunny day, another crew of five (myself among them) with another sailboat departed from a marina in New York City. We were in high spirits, heading right past the Statue of Liberty and then off into the Atlantic with the goal of reaching Bermuda.

The dynamics of what happened on this journey can only be understood by including more about the makeup of the crew. As in teams assembled for corporate innovation purposes (such as with the firm Ideo) that the ongoing mix of social and contextual conditions becomes highly complex.

Offshore Sailing, the company which organized the plan for the New York City to Bermuda trip, was experienced in these offshore plans. In our case, they had an already-tested sailboat, an expert offshore skipper, and a well-equipped vessel including sextants and other equipment for learning to navigate by the stars, moon, and sun. Yes! The initial plan was exciting and the weather looked great.

However! Just hours before departure time, a new less-experienced skipper was flown in from Texas. Also last-minute was the installation of a new electricity generator—crucial to keep batteries powered for navigation and communication technologies and for lights to keep the vessel visible at night. The actual mix of conditions was being rapidly reformulated, and we paying adventurers were kept waiting at the dock for several hours.

The team of five that finally set out for Bermuda a few hours before sunset was far less ideal than the team I described above for the Tortola to Virginia trip. What matters are the actual dynamics of performance as events arise on the ocean, but before we cover such events, here are the oddities of our team. Our skipper was a substitute flown in because an emergency at home required our planned skipper. This skipper was tired, grouchy, and unexperienced with this particular vessel. Myself and one other sailor already had offshore experience and sought to deepen and build on those prior experiences. The other two team members were newlyweds fondly looking forward to the voyage and to Bermuda, but they had only modest sailing skills.

We sailed the first night through the second night under decent winds and among changing but modest wave conditions. Then the chaos began! The dynamics of negative snowballs put us in harm's way. In short, these are the key components that fed negative dynamics:

- New generator failed, so we were approaching no remaining battery power
- Gale-force winds drove waves reaching heights of ten-to-twelve feet
- High waves slowed our progress
- High waves limited visibility of big ships in the shipping lanes we could not yet escape
- Skipper got seasick
- Newlyweds got seasick
- Skipper and newlyweds showed some panic
- Just myself and one other guy shared time steering the boat through the high winds and waves
- Skipper resisted changing our course to a safer one

As I explain below, this maelstrom of complicating conditions, this new chaotic mix, facilitated a personal re-mix of my role, reactions, attentiveness, and deep engagement, and before long a *soaring* experience of coping successfully with the dangerous storm that threatened our lives. As weather conditions grew worse and the psychological conditions of three team members spiraled downward, we finally reached a consensus that we would abandon Bermuda as the goal. Instead, we would aim for safety out of the storm within a day or two at the New Jersey shore.

## My Soaring Experiences Despite the Dangerous Gale

As I steered the boat heading down into a fifteen-foot valley of water, we gathered excellent speed and angled up onto the breast of the next onrushing wave, securing momentum and steerage smoothly enough. Yet, there was no avoiding the highest crest of ocean water ahead, soon pouring over our decks and briefly flooding the cockpit and sousing me head to toe. So as the saltwater shower loomed, I cheerfully bowed my head and let my broad-rimmed San Francisco funky leather hat shield my eyes and nose. Counting one, two, three, four . . . whoosh! My eyes and hat brim surface, I breathe a loud dolphin breath, and the sailboat and I surge forward in unison yet again. Reset—the large waves keep coming, so with dynamic adjustments as acting skipper I do this again and again. Re-mix after re-mix. Dynamic tricky mixing for sure.

Remarkable conditions came together in this gale to support a positive dynamic tricky mix (DTM) for my performance which expanded my sailing tool kit for future sailing in complex seas.

## Survival versus Death on a Summer Day

Here we consider the contrasting dynamic mixes underlying two boating excursions on one summer day on the East River bays off Manhattan Island, New York City.

### Sudden Summer Squall and Tragic Death

Imagine the following chain of events:

- A lone boater with no swimming skills sets out in a rowboat
- The lone boater can row his boat, but with very low speed even in calm water
- A squall is seen approaching
- The boater dismisses the risk of the squall and makes no move to return to shore
- The squall hits at high speed with torrential rain, high winds, thunder, and lightning
- High waves are formed after just a few minutes
- The boater has no life vest
- The rowboat capsizes and throws the boater into the chaotic water
- The boat is tossed far beyond the boater so that it becomes unavailable as a life-saving float

The convergence of these negative dynamic tricky conditions led to death by drowning for this boater.

## Sudden Summer Squall and Soaring Performance by a Racing Team to Escape Danger

On the same day just reviewed, and on the same bay of the East River, I was aboard an Olympic-class racing yacht for advanced racing lessons. The winds, for a while, were perfect, the sun was out, and the lessons were off to a good start. What happened next involved some conditions that overlapped with the tragedy just described, but with the dynamic convergence of other conditions which led to a totally different overall tricky mix.

- A squall is seen approaching
- The squall hits at high speed with torrential rain, high winds, thunder, and lightning
- High waves are formed after just a few minutes with these crucial new components below converging

*Flowing or soaring performances by the sailors in response:*

- The boat crew quickly assesses the high risk of the squall
- The boat crew quickly formulates a strategy to return to the safety of shore
- All crew members have solid sailing experience and boat-handling skills
- The captain/instructor, who is especially skilled, takes charge and issues timely and precise boat maneuver commands
- Each member of the crew assumes a clear role and executes each command with skill and close coordination with the movements/actions of the other crew members
- Each person also has the safety assurance of a high-quality, buoyant life vest

You will recall from earlier detailed chapters of dynamic mix processes that the intensity levels of the components matter greatly, along with the actual set/list of components themselves. In the event of this racing team, the gale unexpectedly threw them into a very important race indeed—a race for their lives toward a safe dock!

You, lucky reader, get a chance to experience the pace and nature of what the sailors were doing in this crisis. Set up two similar chairs and place them four feet apart. Then set the timer on your cellphone to sound an alarm every thirty seconds until this exercise is over for you. Keep this *Breakthroughs* book—physical or digital—secure and where you can read as you move. Good. Now, start reading at this point and continue, but each time the alarm goes off, scramble over to the other chair. Keep your reading going like the boat slicing through the stormy water—but jump to a new position whenever the "captain" alarm gives an order.

The race required the sailors to pull off a series of rapid maneuvers because the only route into the dock was through a narrow channel only six times as wide the length of their boat. The wind was against them, coming from the direction of the dock, so progress was made by "tacking" upwind. The captain called out when to change the boat path from one angle to another.

Cycles on the racing boat similarly repeated again and again—everyone is doing their part to set the sails, rudder, and lines, to jump/scramble over to the other side of the moving boat, and get speed going on the new path. There's just a thirty-second lapse until direction change is ordered again to avoid crashing the boat into the wall of the narrow channel.

Repeat, repeat, repeat. There was very little margin for error. In consequence, the intensity of our attentional focus, our bodily awareness, our planning of each move, our full emotional engagement, our communication, and related factors were high.

In the end, we kept *soaring* along for the ten minutes it took to escape the open and chaotic waters of the bay and reach the safe dock. Without a doubt, this was the most thrilling ten-minute stretch of sailing I have ever experienced.

## Reflections Across Varied Sailing Adventures

We need to revisit the axiom from the beginning of this chapter. Regarding Magellan's Axiom it is evident that, in ocean sailing fifty miles or more from any land, you as a sailor will have no choice but to make any needed adjustments to events that arise. Multiple instances were experienced above.

"When you enter with openness any context where you will often be in 'unfamiliar waters,' the dynamics will be highly favorable for genuine discoveries."

In many instances, conscious choice may be involved. That is, a client, a teacher, a student, or a team may decide to enter unfamiliar territory but always with the option of retreating if difficulties arise. From the viewpoint of dynamic tricky mixes, the most powerful convergent dynamics are favored when the conditions include both high openness and high intensity. These factors will tend to support rich attentiveness and active persistence in the face of challenges and uncertainty. High intensity in response to conditions will lead to large, multiplicative effects on the total force exerted by a tricky mix. Exceptional sensitivity to the ongoing event stream and exceptional and rapid adjustments to the unfolding dynamics, if they converge well with other conditions, will facilitate high achievements of goals and the making of important discoveries.

Let's hear next from an early sailing explorer and a modern poet:

"The sea is dangerous and its storms terrible, but these obstacles have never been sufficient reason to remain ashore . . . Unlike the mediocre, intrepid spirits seek victory over those things that seem impossible . . . It is with an iron will that they embark on the most daring of all endeavors . . . to meet the shadowy future without fear and conquer the unknown."

—Ferdinand Magellan

"When you hear, a mile away and still out of sight, the churn of the water as it begins to swirl and roil, fretting around the sharp rocks—when you hear that unmistakable pounding—when you feel the mist on your mouth and sense ahead the embattlement, the long falls plunging and steaming—then row, row for your life toward it."

—Mary Oliver, *West Wind*

## A Brief Nod to Sweet, Sensuous, Easy Sailing Days

Slow sails in beautiful surroundings involve all the familiar dynamic system processes but yield some experiences that contrast with the high pace and obvious risks in many of the sailing examples so far.

Among the coastal islands of Maine, I have spent many lovely days with friends. On one occasion, in just a few leisurely hours we worked our way through more than thirty small islands. Each island had its own rugged display of rocks, trees, and small inlets. Dolphins relished arching alongside our sailboat, at times riding the wake of the boat, and at other points racing ahead and playing a sort of tag with each other. Seals basked on rocks but also dove in and swam with us as if we were another (rather large) seal in the water. Herons, loons, gulls, and eagles observed our progress. In these circumstances, a "snowball" usually formed to propel our embodied states toward deep relaxation yet keen visual awareness, plus gratitude, delight, mindfulness and a great connectivity to each other.

# Chapter 19

# STARTLING CONTRASTS: MOSTLY STUCK VS. ZOOMING ALONG

Oftentimes situations which we have seen frequently screen our thinking, leading to conclusions that block our noticing—let alone understanding—clear-cut examples that do not fit with our well-established conclusions. Many ignored but illuminating examples, in short, may be right under our noses.

A case in point is learning a second language.

Most adults in the US have seen themselves and their children struggle with a second language in school, failing to reach significant fluency after three-hundred-twenty to six-hundred-forty hours or so of instruction during middle school, high school, or university. Routine conclusion #1: learning a second language is tough for our brains compared to what seems like "easy" rapid mastery of English during the preschool years. Routine conclusion #2: I am just not the kind of person who can master a second language.

Yet, both of these conclusions are questionable because other examples, although less frequent, are easy to find that show certain mixes of learning conditions can make learning a second language easy. These counterexamples have been lying around right under our noses for a long time, yet most laypersons and most language scientists have ignored them.

Excellent dynamic mixes for second language conversations have been shown to lead to high fluency by children and adolescent learners on the basis of about 4 hours per week over fifty weeks. In short, just two hundred hours of these mixes are far more effective than exposure to the three-hundred-twenty to six-hundred-forty hours typical in school instruction.

Such highly positive mixes have involved many different sorts of conversational partners—babysitters, aunts, grandparents, peers, and so on. One fundamental is that, whoever they are, the person acting as a partner is already highly fluent in the target language so the learner repeatedly encounters high challenges and complex sentence examples from the beginning. The rest of the successful dynamic mix, beyond high challenges, we have already seen in other contexts. The conversational partner must truly enjoy and focus on the conversational exchange, follow the learner's leads and interests, bring in positive emotion and humor, show responsiveness to the central messages of the learner, and use specific supportive strategies such as timely questions and recasts/elaborations. That's a pretty long list of cooperating learning conditions, but in practice this highly positive dynamic tricky mix is enjoyable rather than burdensome for the conversational partner.

Right under our noses, the kind of positive and successful dynamic mix for L2 learning just described has supported a myriad of L2 learners. Swedish sign language learning by hearing families of a deaf Swedish child, kids with English L1 learning Spanish or French, kids with Dutch L1 learning English or German, and French Canadian kids learning English as L2 . . . For all these learners and many more similar ones, the proof is in the pudding.

Kids and adults who flourish in second language learning demonstrate that their brains are not "turned off" or "poorly tuned" when it comes to language learning. And the contrast between the typically lousy second language learning in schools and excellent informal conversation learning with the specific mixes described establishes a strong conclusion—whether learning is slow or fast depends upon the richness of dynamic mixes rather than upon the readiness or brain type of the learner. Similarly, in the case of travel, going to a country where the new language for the learner is spoken everywhere is no automatic guarantee; travel contexts only sometimes offer up the rich mixes that matter for explosive progress by a second language learner.

In the rest of this chapter, we present more brief snapshots of what is often hidden in plain sight. What's hidden, in each case, is the high potential for change, progress, and even breakthroughs—hidden because the usual set of conditions and the usual lackluster rate of progress are so inadequate and uninspiring.

For these particular cases, I know both from review of scientific work and from my own experience and those of close friends and relatives how bleak the usual rates of progress can be.

## Mostly Wasted Classroom French and Spanish vs. Fluency After Only One Year of Exposure

I would bet that many of you know firsthand what it is like being in a high school or college class where the "opportunity" to learn a new language is fundamentally wrong. After a few weeks in a course, it is painfully obvious that you and most fellow students are disliking the experience, have made little progress, and have developed extremely low expectations for future improvements. Sure enough, by the end of the year you have probably passed many quizzes but are radically short of language skills that would serve you well in travels to France, Spain, and so on.

Comedians have parodied this situation. Here are a few excerpts from "The Five-Minute University" comedy skit capturing the feeling of many foreign language students.

> Father Guido Sarducci: I gotta this idea for a school I would like to start. Something called the Five-Minute University. And it— the idea is that in five minutes you learn what the average college graduate remembers five years after he or she is out of-a school. It would cost like twenty dollars. That might seem like a lot of money, twenty dollars for just five minutes, but that's for like tuition, cap and gown rental, graduation picture, snacks, everything. Everything included.
>
> In Spanish, students learned only "¿Como esta usted?" and "Muy bien." In economics, students learned only supply and demand.

In any situation where you might like to see astonishing rates of skills acquisition, here is a good question to ask: Who else has achieved this, and how?

For rapid fluency in a new language, interesting answers lie in the histories of children who are—pleasantly, for the most part—uprooted from their local neighborhoods and schools and trotted off with their parents for a year in a totally unfamiliar country with its own contrasting language and multiple contrasting cultural patterns.

One case history I know well because it concerns my close friend and colleague Mikael Heimann. When he first came to study developmental psychology with me, he brought his Swedish family to State College, Pennsylvania. His daughter Anni entered here a grade five classroom where she was the only Swedish speaker. Her English skills were nil. Yet, she was accepted by the other children and teachers at the school and entered rich and emotionally engaging interactions every school day. After eight months in this totally new mix of conditions for language, her English had leaped to the level of the other children. Wow! In one year of sink-or-swim social conditions she had gained nine years of English spoken language. So in a typical week she had been *soaring* along in her new language. This just kept going and, in all likelihood, was snowballing over time to faster and faster language-learning rates.

Likewise, in more instances than usually acknowledged in linguistics and psychology, adults under rich dynamic mixes prove capable of gaining fluency in a new language in just one-to-two years. Here we give nutshell accounts of some individual cases and some groups of highly successful adult language learners.

*Love and Language Case*. Romanian woman meets Swedish man. The woman knows zero Swedish, but in this continuing love relationship gains Swedish fluency. Similarly, the Swedish man goes from zero to Romanian fluency. In Dynamic Tricky Mix terms, mastery of a new language occurs for such couples as follows. (1) A truly fluent conversational partner is available, (2) the conversations are highly positive emotionally, (3) there is mutual attraction and love, (4) there are new professional benefits for learning, (5) the conversations are frequent, wide-ranging, and across many varied contexts, (6) deepening expertise in the new language further enhances intimacy and relationship satisfaction, (7) gaining modest levels of mastery snowballs into extremely rich new conversations and acquisition of the highest complexity of structures in a language.

*Enhanced Military/Political Advantage, Case 1.* An American Midwestern man graduates from an extremely small high school with no second language offerings and no other opportunities to pick up a second language. For context, it's 1961 with the Cold War Russian/American conflicts and competition. This close friend of mine then enters an Air Force intensive language program for two years. Believe me, his level of Russian language expertise was intensely evaluated, and he showed brilliant performance. With these skills, he was able to participate in many government-sponsored activities. As just one impressive example, he flew in American jets, monitored ongoing communications between Russian pilots and navigators, and in some instances tricked the Russians into believing that he was a Russian in another plane.

*Other Enhanced Military/Political Advantages.* In similar cases to the one above, military or governmental authorities have at times trained skilled individuals in the language of newly emerging competitors or perceived enemies. Many countries have created new dynamic mixes which give financial support toward language learning, intensive and highly interactive language-learning experiences, group bonding, and high incentives for gaining fluency in a new target language. Computer code engineers, pilots, data analysts, and myriad other professionals have shown high success in language mastery programs set up by Israel, the USA, Iran, China, Russia, and others.

Again and again, such adults document that an individual who has been *stuck* in making any progress in any new language for years will easily shift into *flowing* or *soaring* rapid-acquisition modes when dramatically new and rich mixes of supportive convergent conditions are arranged.

## Advances Over Several Decades in Clinical Treatment of Two Very Different Disorders: Adult Anxious Phobias and Severe Language Delay in Four-to Seven-Year-Olds

How do we know that clinical treatments are effective at all, let alone highly effective? Ideally, by accumulating careful, rigorous evaluation of outcomes carried out by many various research teams covering many different clients and treatment agents/clinicians, and discovering that positive effects are consistently occurring and documenting how strong the gains are.

That kind of clear evidence was *lacking* around the years 1965–1970 both for phobias and childhood language delay.

Yet, by about the year 2000, it pleases me to review that excellent, highly effective treatment approaches for both disorders were well documented. In both cases, these advances were generated through a series of creative, successive re-mixings of treatment conditions which were shared across clinicians and researchers and evaluated rigorously by multiple teams.

At the same time, the particulars of effective treatment mixes show some fairly dramatic differences.

Phobic responses to snakes, elevators, dogs, and more proved to be decreased or even fully resolved by therapists guiding the client through examples/images in a graded, systematic series from the least-threatening examples (e.g., one snake photo) to increasingly realistic and direct encounters (e.g., the client holding a snake in their hands). In dynamic tricky mix terms, the client at each successive step learned to relax in the presence of each example, to carry into the next encounters a new expectation of coping well, to demonstrate socially that they are changing in a positive way, to extinguish strong links between fear/anxiety and the phobic object, and to acquire new and interesting information (snakes are not slimy, they do fascinating things, etc.). Classic work by Albert Bandura, Tom Borkovec, and others used Cognitive-Behavioral Theories to inspire this work.

In the case of children's language delay, in Chapter 14 a very different pattern was observed. Behavioral Theories initially had led to reliance on repeating specific language forms after an adult model—talk*ing*, walk*ing*, sing*ing*, for example, were laid out for imitation by the child. However, a far more effective treatment approach was developed in which, during lively, ongoing conversation with a child, the therapists worked in a new kind of tricky mix. In this "conversational recast intervention" children readily processed and learned new language structures because the "uptake" by the child was supported by a rich mix of positive emotions and expectations, an absorption in the meaning of the conversation, and contrasts between what the child has just expressed and a responsive but challenging reply by the adult.

As research continued on adult anxiety disorders as well as child language disorders, it proved possible to create refined tricky mixes which enabled clinicians to use the general approaches as just described but with adjustments in the mix to the characteristics of different clients (e.g., Newman et al., 2008).

## Four-Year-Olds Seen as Cognitively Far Below Adults vs. Kids' Cognitive "Computers" Are Actually Raring To Go and Just Waiting for Challenging Mixes

In psychological cognitive tests, the information-processing skills of four-year-olds often appear to be very weak compared to those of adults. Their brains might therefore seem unprepared for complex challenges, and certainly not prepared to handle the kinds of content typical of first grade and second grade classrooms. Yet, because this is widely believed, the kinds of cognitive tests actually given are often severely restricted in complexity for four-year-olds.

For example, working memory might be tested by the examiner reading the sequence "four, eight, one, three, nine, five" and then asking the child to repeat those digits. Typical spontaneous performance of a child might be just three or four of these digits.

### In Contrast, New Test Mixes Show Four-Year-Olds Match or Surpass Adults in Handling Complex Rapid Information Flows

An important and informative method for testing processing of rapid visual information is termed "RSVP"—rapid serial visual presentation. If you are shown a rapid series of pictures where each picture is shown for less than a second, you will find it impossible to place all of the items into longer term memory. Still, like other adults, you probably will be able to process some of the pictures and later recognize them. Now the question at hand is how well can four-year-olds do with the same pictures under exactly the same conditions of rapid exposures? To our surprise and delight, the four-year-olds rise up to the adult level of memory performance.

These excellent processing capacities of young children help to make sense of the clear successes in learning new concepts, new words, and new sentence structures by children all over the world (preschool or not) during the period between two and five years of age.

## Alison Gopnik's Observations

Gopnik has done experiments and made other observations that also emphasize the cognitive strengths rather than weaknesses of young children. For example, children often show fascination with the fine details of the sensory world that adults ignore while pursuing focused goals. The children, in a way, are truly "soaring" in their open exploration of the world around them. Then, for certain problem-solving situations, their openness leads to more rapid shifts in strategies and quicker discovery of how a toy's movements and lights are triggered than adults can achieve.

> "There are children playing in the street who could solve some of my top problems in physics because they have modes of sensory perception that I lost long ago."
>
> —J. Robert Oppenheimer

## With Novel 3D Lego Sculptures Four-Year-Olds Match Adult Abilities in Spotting Similarities Between Sculptures

Pattern detection processes are at the heart of learning all kinds of complex skills—language, mathematics, chess, reading, art, and sculpture. Research rarely makes direct comparison between the performances of adults and preschool children for a novel set of patterns.

In my first teaching position after receiving my PhD at Yale, I was fortunate to work with two talented graduate students at Stanford University. Stephen Kosslyn and Phipps Arabie and I constructed multiple Lego sculptures and compared in a complex analysis the "multidimensional space" of how these sculptures were perceived. It turned out that what four-year-olds and adults saw as highly similar, moderately similar, quite different, and so on, showed remarkably few contrasts despite the age differences. In short, our conclusion was that the visual 3D pattern detection and comparison processes of four-year-olds were already highly sophisticated.

The central breakthrough in all of the above examples is that three-to-four-year-olds have far more advanced brains and processing capacities than we usually credit them. In turn, that change in how we view young children should spark us to create new breakthroughs in highly engaging and highly stimulating learning environments for young children.

## Down Syndrome Leads Individuals to Lifelong Restrictions vs. Integration into Family and Community Opportunities

When I was in graduate school at Yale, we took some field trips to New Haven Connecticut to see how the state of Connecticut handled individuals diagnosed with disabilities including Down Syndrome. It was a sobering experience.

For example, Down Syndrome children and adults in 1968 led very mundane lives. Little was expected of them so they engaged in simple activities of object manipulation (assembling flashlights, etc.), watched TV, or just sat around. Compared to the "bad, old days" of asylums and other institutions for those with disabilities, the "campus" in Connecticut had nice open spaces, clean rooms, and their caretakers were attentive. But the procedures in place guaranteed that they would not be given learning opportunities for complex social skills or complex language, let alone any literacy or academic challenges.

Low expectations of children with Down Syndrome have long been in place. In fact, in antiquity, many infants with disabilities were either killed or abandoned. Yet, at other times there are hints of some acceptance and support. For example, some quite old instances of art are believed to portray Down Syndrome, including pottery from the pre-Columbian Tumaco-La Tolita culture and the 16th-century painting *The Adoration of the Christ Child*.

By 1974, our lab had breakthrough demonstrations showing that, for typical preschool children, language progress could be greatly accelerated by causal experiments introducing conversations with high challenges mixed with converging highly positive emotion and engagement. We decided to put aside the usual expectations of Down Syndrome children and conduct a similar causal experiment. It worked! Elsa Stella-Prorok and I showed that pre-intervention children were essentially *stuck* in language progress but shifted into higher modes of learning when rich conversations were created. Complex levels of language never before seen for such children were demonstrated in our study. Clearly, the Down Syndrome children had readiness cognitively for complex language all along, but only moved forward when new mixes of interaction were introduced. As other projects around the world gradually materialized, general expectations of Down Syndrome children and adults were transformed for the better.

So these days it is common to see that social, career, and family status is far above the levels of one hundred years ago. Recognition of their performances as actors and in other pursuits is widespread for a number of modern individuals with Down Syndrome—another indication that re-mixed expectations and opportunities open up some remarkable achievements.

Here are the names and activity areas for just a few of these modern individuals who have achieved considerable recognition: Karen Gaffney (disability activist), Sarah Gordy (actress), Pascal Duquenne (actor), and Collette Divitto (cookie company entrepreneur).

## Creating Zooming Conditions for Children Stuck in Their Math Progress

We earlier noted that schools that tried to choose the "single best way" of instructing in mathematics often ended up leaving a subset of children far behind. For children who have made *no* clear progress across grade one to grade five, it may seem reasonable to assume that these kids just do not have a "math brain." That conclusion is totally refuted when an effective tutor works with one of the fifth graders a couple of times a week over four months and that child jumps four-to-five years in their math skills. The child did not need a new brain—instead, what was needed was a fresh dynamic tricky mix of learning conditions. Mark Lepper and his colleagues demonstrated this in their research in 1993.

Their research showed that only certain tutors created these strong "soaring" or "zooming" rates of progress. Key aspects of the powerful tutors were these: humor, warmth, expressions of confidence in the child, and when a child failed a problem the tutor assumed responsibility while all successes were attributed to the child's learning ability and their efforts. That is a tricky combination to create and maintain, and only certain tutors pulled it off. In other words, "give the child a tutor" is too simplistic an approach—instead, the complex details of effective tutoring are essential to keep in mind.

# Chapter 20

# WHEN LABELS BECOME SACRED COWS

L abels of all kinds can cut many ways. They often help us quickly zone in on a field of meaning but carry the risk of blinding us to differentiated perception and understanding.

Many labels are so overused and so inappropriate that we must treat them as sacred cows that get retired to a remote pasture.

Just notice what comes to mind when you read each of the following labels: "Dyslexic," "Working Class," "Visual Cortex," "Welfare," "Conservative," "Liberal," "Gifted," "Evolution," "Injustice," "Boss," "Pollution" . . . In each case, what you dynamically activate as meanings will be somewhat different than my meanings, and we should take care to dialogue enough to arrive at more precise understandings. But for the moment, just note that the labels evoke in you a complex mix of facts, personal history and responses, attitudes, expectations, evaluations, and so on and not just a simple dictionary-like meaning.

## Educational Innovations to Boost Literacy for Children with Autism

Here's another label: "Autistic." Give that label to a child and the risk is that we may expect to see the same characteristics as what we think we know of every other "autistic" child. And therefore we become at risk of thinking we know that this child with some autistic symptoms should be approached in a standard way designed for "the autistic child."

As part of my lab's research, we have designed highly animated software lessons that may help children with autism or children any place on the autism spectrum improve their reading and writing. The insanity of expecting the autism label to capture the range of actual behaviors of children was driven home for me when I visited a school in Sweden for children with autism diagnoses. The children there involved in our literacy-boosting research all began their new lessons with a history of very low literacy with evident dyslexia.

"Eric" was very social toward me and toward his teacher as he sat down for a lesson on astronauts, dinosaurs, and more in the reading/writing program. In this context he was not socially "autistic" and he was ready to attend to the computer and his teacher and went right to work. Surprisingly, though, a hurdle came up for Eric that was cognitive—he had trouble remembering the relations between videos and the messages in text he needed to master. So moving toward an effective dynamic mix for him was not about social/emotional processes; instead, it required "tweaking" aspects of how the computer material was timed and presented so that it supported his processing profile.

In contrast, "Sophia" was a handful to engage. Socially, she showed the classic autistic tendencies of self-absorption, restless motor movement, and poor use of eye contact and language to socially cooperate with others. In consequence, her teacher needed to make "tricky" move after move to bring her back into the social interaction and to cooperatively manage the computer interface so that Sophia would take in what the computer displayed. *If* she maintained attention, she had a great chance to process what the computer screen flashed up—a sentence in Swedish text to be learned, Swedish spoken language, and a video that exactly matched the meaning. It was indeed tricky, but her teacher was skilled and patient, and Sophia became part of the dynamic tricky mix we were after. That is, during her brief attentive/engaged episodes she was socially cooperative and actively processed all the relevant material and made progress in learning how to read and write sentences. We could demonstrate this achievement through her excellent performance on new computer episodes that required her to write/input correct text on test videos where the video comes first and she filled in the Swedish text appropriate to each video. For her, the effective dynamic mixes were interspersed with brief resistive, distracted episodes, but those more negative episodes did not prevent the progress in literacy she achieved with the interspersed positive mixes.

## Rehab in Cambodian Villages for Amputees: Sacred Cow Labels of "Amputee," "Prosthesis," and "Western Medicine"

In 1981, a young neurologist by the name of Joe Julian from the United Stated flew for a one-year stint to help in Cambodia after the end of the Cambodian Civil War in 1975. There had been killings or the severe wounding of many doctors and nurses by the victorious Khmer Rouge. Joe became a specialist in trying to provide effective rehab for the many amputees created by exploding land mines.

I am blessed to currently have Joe as a friend and intellectual collaborator here at Penn State University, so I am very pleased to fit some of his experiences into this book.

Before long, Joe discovered that he needed to do a lot of rethinking. The very labels "Western medicine," "amputee," and "prosthesis" took on new meanings in the context of these Cambodian villages near the border of Thailand. Western medicine and his role in medicine were perceived as suspicious intrusions into the local culture, even if the doctor and his staff are trying to provide services that were in no other way available to the villagers. Similarly, a prosthesis made in a factory with sophisticated metals and plastics back in the Western world might indeed be sophisticated and, if worn, a very effective limb. But all efforts to help might come to naught if a particular amputee found these imported prostheses too strange and inappropriate for wearing back in the villages (even after considerable effort and time had been expended toward providing the right fit, training, and rehab exercises to that person). The label "amputee," in this context, does not always carry with it the meaning that this person was now missing a limb which he or she fervently desired to have replaced with the best artificial limb that could be found.

Joe began an ambitious re-mixing of the materials used in prostheses and the ways in which he and his staff tried to achieve what they considered appropriate care in the context of this war-torn society and its culture. Bamboo, which is readily and cheaply available locally, was incorporated into artificial limbs. Workers who are skilled at harvesting and working with bamboo are also readily available. So, as an experiment, fairly primitive-looking artificial limbs that combine leather, bamboo, and other local materials were created. Even the first of these innovations worked fairly well. These creations were more acceptable to the wearer. When a prosthesis was taken back by a client to the village, the client was much more likely to continue using it and engage in a range of activities that were enabled by an effective artificial limb.

Joe Julian also dynamically re-mixed his role to become part of a culturally sensitive team rather than the "expert," take-charge leader and commander of his staff. In his words, he became "a facilitator and communicator for the work and wisdom of others instead of trying to do it all [himself]." His team's work continued to evolve when Joe stayed in Cambodia beyond that initial year. He was able to bring in many apprentices to his workshop and to teach them to be effective designers and coaches for new prostheses for the amputees. Joe and his teams were employing very clearly those aspects of an effective, evolving, positive dynamic tricky mix that we have so often stressed in this book. They were trying bold and creative experimental variations of design and procedure, and they were also engaged in active and continuous exploration, improvisation, monitoring, and readjustment and re-mixing of what they provided to each individual client. In effect, in the war-torn circumstances they were in, they found that they could achieve more effective change with less sophisticated materials. This approach was productive as long as they persisted in learning from trial and error and experimental variations and maintained continual work despite many mistakes along the way.

Joe's experiences in Cambodia early in his career led to changes that fit with the emphasis we have placed on the power of key complex self-changes that carry forward into new situations, new problems, new teams of people. His work in Cambodia transformed his core self-concept and his core philosophy. Joe describes components of this core philosophy this way: "I became an aware, conscientious steward of resources . . . I shifted from valuing facts to valuing understanding, which meant respecting the opinions, concerns, and values of the whole therapeutic community of staff, patients, families, and more." Joe has seen that this new core philosophy formed in Cambodia has persisted across decades to influence and inform not only his professional life but also his personal life.

What Joe achieved in Cambodia also ended up illustrating well the distinction we've made many times between truly impossible events and events that are merely improbable until dramatic re-mixes occur. Nearly all rehab professionals in the West believed that new and effective staff could not be trained without a long training period. Nevertheless, working within his new, open, community-sensitive philosophical framework, Joe was able to pull off rapid, remarkable training successes in his clinics and workshops. He trained new rehab nurses, physical and occupational therapists, and designers and engineers, all with no previous medical or rehab background. Remarkable! Yet, this is fully consistent with our repeated discovery that under dramatically new, intense, convergent, and bold dynamic tricky mixes surprising breakthroughs become not just possible, but often predictable.

## Seeking Dramatic Advances in All Educational Domains for Children *Stuck* in Special Education

Between the years 1980 and 2000, a story emerged of shifting sacred cows and their impacts. It is a revealing story about the danger of labels and also the danger of the simplistic, flawed strategy we examined earlier—"Choose the best strategy and reject all other strategies."

In the United States and in many countries abroad, children who had been *stuck* in their educational progress were placed in totally separate classes, and sometimes in totally separate schools. Among those students, many held the limiting labels of "developmental delay," "severely learning disabled," "dyslexic," or "autistic." "Special education" is itself a potential sacred cow label for the set of procedures and expectations most often occurring in those contexts. The approach of teachers in these contexts also was influenced by the low expectations for anyone with the four labels I just mentioned above. After children have been resident for instruction in the special education context for many months or years, as is typical, they continue to be *stuck* at a minimal or zero rate of learning. Then negative snowballs are created that tend to make the probability of improved educational outcomes extremely low. Among the mix that creates negative converging dynamics are these factors: lessons with minimal challenges, low expectations of engagement and learning by both children and teachers, failures to explore and experiment with fundamentally new procedures that carry higher levels of challenges, and limited awareness of any breakthroughs in the science of learning and innovations in education that respect those new insights.

## New Sacred Cow: Facilitated Communication as a Fantastic Breakthrough Which Needs No Critical Examination

Consider the following scenario which became all too frequent during the 1980s and 1990s. Children who had never acquired any significant literacy or mathematics skills were given a new procedure and embedded in ordinary classes. There the children were constantly accompanied by an adult "facilitator." Messages were created by the children entering letters and words on a computer keyboard that then led to output from an attached printer.

Douglas Biklen of Syracuse University and other proponents of this procedure termed it "facilitated communication" (FC). Their key claim was that children with prior histories of severe learning disabilities had typically already acquired a wide range of skills that were not being recognized because the child could not effectively communicate what they had learned. Messages coming from the computer and printer during FC sessions were interpreted as supporting these conclusions. Among the messages from these children who had been *stuck* in their educational and communicative progress were complex text accounts of their own history, their current thoughts, essays and poetry, and complex expectations for the future. Seems almost too good to be true, huh? Well, just read on.

So, what was the mix of conditions that the proponents of facilitated communication *claimed* were creating innovative, breakthrough, miracle-like changes in special education children with terrible learning histories? The facilitating person sits close to the child and supports the child's arm that is posed over the keyboard. Keys are then pressed by the child's index finger. This physical support may create some shift in how easy it would be for the child to enter one letter at a time to make words and sentences. Along with that contact, emotional support and encouragement from the facilitator would also feed into the process. New, high expectations from the teachers and from the children's parents likely arose because the facilitated communication practice was so heavily promoted and because everyone could see that many children were benefitting. CBS News and many other media outlets gave attention and their enthusiastic blessing to FC, unblushingly labeling he child's performances during FC as miracles and breakthroughs.

Scrutiny was strongly resisted by the promoters of facilitated communication. Thousands of children around the world were paired with personal facilitators. With those facilitators, the children regularly attended a broad range of academically challenging classes. Note again that the messages attributed to these children were those that came out of printers after being created one letter at a time by hunt-and-peck typing by the child with physical contact by the facilitator throughout. It was claimed that the children now demonstrated reading, writing, and mathematical calculation as well as analysis of literature and scientific experiments that they had in no way ever demonstrated before. Part of the tricky mix of factors that kept facilitating communication procedures spreading rapidly was that, of course, teachers and parents alike wanted to believe that some miraculous new set of skills and learning capacities had been discovered in their severely learning-disabled children. Hope can be part of a powerful mix that keeps a set of activities going.

A new aspect of this facilitated communication phenomena appeared by examining other messages coming out of the children's computers and printers. Such messages asserted for some children that they had been repeatedly sexually abused by parents and other family members. Without further checking, on the basis of those messages alone, social welfare authorities in multiple states and in multiple instances removed children from their homes and placed them in foster care. Thus, another label that was partially used against these children and their family was the label of "sexually-abused child." That label is very emotionally laden, and it seemed to many to require speedy action to protect the children. Facilitated communication procedures were treated as is if they had uncovered otherwise unknown instances of abuse.

We can sum up this range of events concerning facilitated communication within the framework of Dynamic Systems. A tight cluster or tricky mix combination of dynamic and highly convergent conditions contributed to a negative snowball of more and more widespread use of facilitated communication. This occurred without any of the careful monitoring, experimentation, and adjustment of tricky mixes that we have stressed in this book as essential for strong progress. In the case of FC, these approaches were needed to ensure an accurate account of what is really helping children and what is not helping children.

In the next section, it will be revealed how scientists and others came to know that users of facilitated communication had been living in an illusory world where they believed that a highly positive dynamic snowball of ever-improving knowledge, skills, and learning along with bright prospects for further education was evolving for children who use facilitated communication.

## Revised Perspectives: Healthy Mistrust of All Labels and Procedures, Leading to a Wide Range of Bold Experiments and Monitoring of Those Experiments

If we examine any set of teaching/learning interactions with a critical and open mind, it is usually not very expensive or time-consuming to implement a range of rigorous evaluations. These evaluations need to focus both on the outcomes of learning progress and on the character of the social-emotional interactions.

As soon as rigorous monitoring and evaluation of facilitation communication emerged, troubling findings came to light. If the facilitators were given a blindfold so that they could not see the computer keyboard, then no sensible messages came out of the child/facilitator interactions. Oops! This and other findings raised the possibility that the messages being attributed to special education children were in fact being created in the facilitators' minds and were independent of the children's minds. The messages thus seemed instead to be reflecting knowledge, expectations, and beliefs held by the facilitator.

A series of independent investigations all converged in their central conclusion: the messages coming from the computers and printers during facilitated communication were bogus as indications of thoughts, knowledge, or skills for the children involved. Children, teachers, and facilitators had all been wasting their time!

In other circumstances, strong hopes can serve as components of a highly positive dynamic tricky mix. But within the facilitated communication bandwagon community, false hopes were helping to hold together a toxic mix of actions. Broad recognition of the limitations and dangers of facilitated communication has developed from about 2001–2020. Nevertheless, some websites and media still promote FC today as a powerful and appropriate procedure for children and adults on the autism spectrum.

So, how to proceed toward a re-mix of conditions for those who had been involved in facilitated communication? First, when monitoring shows that an activity is wasting time, stop it immediately. This means that new activities adjusted to individual children should be generated, drawing upon the energies and hopes of teachers, parents, and others who interact with the children. Because the children had been and still were *stuck* in terms of significant acquisition of ordinary skills expected for children of their age, the dynamic tricky mix approach strongly indicates that new experimental variations should be implemented and continually monitored on who is interacting with the child, with what procedures, and in what context . Only activities where good social interactive dynamics and interesting new levels of learning occur should be continued, re-mixed for even further impact, and then monitored and adjusted continuously as needed.

# Chapter 21

# DRACULAS, DUNCES, AND OTHER OBSTACLES TO INNOVATION, CHANGE, AND BREAKTHROUGHS

Sometimes you may find yourself part of a strong, positive, creative new tricky mix which is working well until unanticipated obstacles arise. We explore here a few variations of such obstacles and how they affect ongoing dynamics. In later chapters we will expand on this topic.

## Draculas vs. Muses

Imagine you have a truly great and original idea. You've put in time exploring, dreaming, scarching/researching, and it's paid off. As you forge ahead, remember to take a few clues from others in diverse fields who have gone down these paths before.

Because your idea is fresh, new, and original, it likely will be in a "fragile" state for a while, so choose carefully who you share this idea with.

In particular, beware of any folks who have a history of negativism, skepticism, and inaction. These are Dracula-like and will tend to "suck" the enthusiasm out of an innovator like you. "No way," they will say. "Don't get your hopes up," they will say. These responses are not based upon insight or special knowledge but are habitual ways of responding to anyone else's initiatives and originality.

Julia Cameron in *The Artist's Way* describes the urgency of avoiding truly negative people, "Draculas":

"Those who dampen us act above it all. No matter how interesting our thoughts or conjectures, they seem to be barely stifling a yawn. Boredom is the pose they take most often . . . Cynics are monsters to the creative process."

Our progress may also be seriously disrupted by folks who are not as consistently negative as Draculas but who are self-centered and make unreasonable and poorly timed demands of us. Part of the positive dynamics in any highly effective team project is finding tricky mix strategies which bring out the team's most constructive, focused, collaborative, and persistent tendencies. Even one or a few disruptive people can wreck the positive dynamics. Cameron has a term for these—"Crazymakers."

"Muses" are the contrasting individuals who show patience, genuine interest, respect, encouragement, insight, and warmth. They inspire leaders and team members alike. They spark needed experiments, tweaking, bold explorations, and confidence. Muses inspire as many rounds of dynamic re-mixing as prove necessary for achievement of genuine, significant breakthroughs.

## Don't Undercut Your Own Progress

To make substantial innovations, we need to guard not only against external "Draculas" and other negative sources, but we must also guard against our own least constructive tendencies. In both cases, we should repeatedly seek insight into how to give up the "bad habits" of familiar, easy to remember, emotionally invested but fatally flawed and too-simple approaches. Further, our strategic plans should incorporate times/contexts for regularly tuning up our mindfulness, self-awareness, active monitoring, and adjustment of our spirits as well as our action plans.

### Pema Chodron's Ideas on Awareness

Pema Chodron is an inspiring writer from the Buddhist perspective. She captures in her prose how easy it is to feel wounded by others' actions and attitudes and to escalate these temporary wounds into persistent and self-defeating strategies that sap our energy, awareness, and effectiveness. Thus, we can become an enemy to ourselves and our plans that far exceed threats from the outside. Here is one sample.

"When everything falls apart and we feel uncertainty, disappointment, shock, embarrassment . . . we feel the queasiness and uncertainty of being in no-man's-land and enlarge the feeling and march it down the street with banners that proclaim how bad everything is. We knock on every door asking people to sign petitions until there is a whole army of people who agree with us that everything is wrong . . . We could just sit with the emotional energy and let it pass. If we can look at and see the wildness of emotion, we can not only begin to befriend and soften toward ourselves, but we can also begin to befriend all human beings and indeed all living beings . . . We begin to develop true compassion for ourselves and everyone else because we see what happens and how we react when things fall apart. That awareness is what turns the sword into a flower. It is how what is seemingly ugly and problematic and unwanted actually becomes our teacher."

## Julia Cameron on Grounding Processes

In her book The Artist's Way, mentioned previously, Julia Cameron touches on the topic of grounding and where the creative mind emerges from. Her perspective sheds light on the importance of knowing ourselves and how keen attention to both reality and imagination play into innovation, change, and breakthroughs:

"People frequently believe the creative life is grounded in fantasy. The more difficult truth is that creativity is grounded in reality, in the particular, the focused, the well-observed or specifically imagined. As we lose our vagueness about ourselves, our values, our life situation, we become available to the moment. It is there, in the particular, that we contact the creating self . . . We become original because we become something specific: an origin from which work flows."

# Chapter 22

# SNOWBALLS

We have explored many different kinds of complex events. By deliberately and persistently trying to move beyond simplistic approaches, we create better chances of positive outcomes. This holds true for such seemingly unrelated domains as:

- Learning by children with disabilities
- Learning by children who are absolutely ordinary
- Enhancing skills in sports, music, or art
- Biological events and processes
- Social change issues such as climate change, endangered species, and economic equality

Here we consider dynamic processes which create "snowballs" over time. Snowballs are dynamic patterns that grow larger over time. For example, when a child begins to learn in week one and then returns each week for thirty weeks to the same context and teacher, we can track whether whole bundles of learning factors work together dynamically to foster ever more powerful, rapid learning. It is "tricky," as we have seen, to initiate and sustain this kind of snowballing, positive, ever more favorable pattern. Also, by the same dynamic processes, it is not unusual to find that we have unintentionally created the scary opposite—an ever more unfavorable "negative snowball." In this chapter, we consider both sorts of snowballs.

## Positive Snowballs for Children and Teachers

In some fortunate instances, our lab has been able to not only launch rapid learning advances by children with autism, but also have enough people and resources to richly document the learning conditions accompanying the children's successes. The work in Sweden provides a relevant example. The children studied there with autism spectrum disorders were stalled at extremely low skill levels in reading before we introduced a new combination of reading software and a package of teacher strategies to bring in highly positive social-emotional and coaching conditions. Given this new approach, in the first few weeks the children achieved fairly good rates. The snowball really took off then. For a dozen weeks or so, most of the children increased the learning rate every week. Our argument is that this positive snowball included the successful dynamic convergence during reading lessons of at least the following conditions:

- The child sees totally new technology-based parallels between video events, their written meaning, and their spoken meaning
- The child is given freedom to choose most of the events they explore
- No tests occur until fun with exploring and probable literacy advances based on that have taken place
- Short, fun tests make obvious to both child and teacher that new learning has occurred
- The teacher relaxes into a role of encouraging, providing brief conversational exchanges, and praising the child's efforts and achievements
- Events within the software trigger delight, surprise, and humor
- The information-processing opportunities richly support rapid comparisons between video, text, and voice presentations/recasts of the same meaning

In this research, we also had funds to follow up on the children and teachers after we had exhausted all the computer lessons then available. The teachers returned to their usual teaching materials—but with crucial differences we will pinpoint in a moment. The results of the children's reading tests over the following twelve weeks held a very pleasant surprise: the children now were learning far faster than they had *before* they met our special intervention with the computers.

We argue that, even without the high challenges in reading and the particular ways that the computer could lay out text sentences, matching video, and matching audio, children learned because the mix of conditions after intervention that now occurred in the classroom had shifted in crucial ways. Teachers and kids alike knew that the kids had reached new literacy knowledge and that more learning was a realistic expectation. The teachers, accordingly, were more active and more experimental in what they tried out in their teaching, and the children brought into their lessons heightened engagement, attention, and initiative.

In short, a good snowball of learning kept going when the usual classroom now included the transformed teachers and children with enhanced confidence, positive emotion, and raised expectations of productive lessons and fun along with further learning successes. This is another illustration of how the discussions in Chapter 17 about "complex changes in self" as powerful catalysts toward breakthroughs can be at play in a varied range of contexts.

## Snowballs in Corporate Settings

When we think about significant shifts over time in corporate settings, we need to think about how individuals are changing as well as how the overall corporate "beast" is evolving. Consider now a few case examples that help illustrate how dynamic tricky mixes apply.

### "Simple" Products Have Complex Meanings

If choosing a bank for simple financial services were all that consumers considered, those banks with the best objective services would thrive the most. Ditto for buying a car based on objective performance and appearance values.

Instead, there are some dramatic illustrations that consumers hope for, and they base their banking decisions upon more complex mixes that incorporate perceived fit with the consumer's personal identity and lifestyle. When Pat Fallon and his marketing firm set out to create a new advertising approach, Fallon first sought to identify the characteristics of the bank's likely customers. As reviewed in the book *Juicing the Orange*, a key concept emerged—that these customers were "balance seekers." They value money as a facilitator of things they want to do, but not as a predominant goal.

Citibank accepted Fallon's approach, with multiple variations on posters, TV ads, and so on that stressed there is more to life than money, Citibank understands that, and Citibank will help you live richly. Fallon viewed these as platform ideas that could cascade into all aspects of the bank's business and its relationships with customers.

Positive snowball effects ensued. Interest in and trust in Citibank rose. Credit card holders greatly expanded. Loan applications rose both from businesses and individuals. Positive customer experiences were shared with still further potential customers through social media. Bank employees raised their expectations and took new actions in line with the platform ideas, all of which further fed into positive customer experiences, expectations, and choices of Citibank's products in preference to those offered by other financial institutions.

## A Golf Headline Interpreted within Snowballing Mixes

Can chewing gum transform a golfer's game? This possibility showed up as a headline concerning Jason Spieth and his approach to the British Open in 2017.

Yes! The gum chewing could be one component of an increasingly powerful, positive dynamic golf snowball. But this is tricky, complex, and potentially elusive.

On the plus side, there are multiple ways this could work. Chewing could lead to changes in certain neurotransmitters and in the release of adrenaline, lowering anxiety and thus helping the golfer retrieve and use the best possible motor patterns to deliver splendid shots again and again.

Chewing on gum could also have an indirect positive effect through raised expectations of high performance which in turn tweak perception, attention, and self-monitoring toward more optimal dynamic mixes.

However, the same processes at work could equally contribute to disastrous performance. If the golfer obsessively chews gum, the initial positive physiological effects may soon turn toward dysregulation and impaired motor patterns. Similarly, if high expectations steer the golfer toward nonchalance and poor attention to fine details of the current context, then poor shots may occur that lead to some "boomerang" effects. For example, the stark contrast between high expectation and low performance may lead to anger, dysregulation overall, and reckless shifts in strategy. That would all add up to a convergence of negative conditions and a dysfunctional snowball.

## More on Spectacular Snowballing Growth Rates

Because we have in earlier chapters explored many instances and aspects of positive dynamic tricky mixes and their facilitation of breakthroughs, we are in a good position to expand into a discussion on spectacular growth rates based upon unusually powerful dynamic mixes.

In a stickleback fish or a seagull, rapid onset of a behavior is sometimes directly facilitated by an inbuilt, instinctual, biologically given template in the brain. However, the corresponding notion of Chomsky that only a powerful innate template could explain how human toddlers are able to progress in the beautiful complexity of spoken language is spectacularly wrong. The last fifty years of investigation into children's language learning shows instead that humans are prepared by the openness of the young brain to master any complex communication system *if* a fundamental set of social, emotional, and cognitive conditions are met. If lacking such cooperating positive dynamic mixes, children—over and above any genetic disposition—are at greater risk of becoming seriously delayed—as in specific language impairment—or almost totally deficient in language—as in severe autism spectrum disorder. These issues get a close look in Chapter 24.

By the same dynamic theoretical framework, we see how easily some children with unusually rich dynamic mixes proceed between one and four years of age to make exceptional or even spectacular language growth. In fact, some children acquire full fluency in four or five spoken languages and other children acquire fluency in sign language and one or two spoken languages. Still other children focus their early language learning mechanisms on written language as well and are fluent in text production and comprehension along with spoken language by age four.

Another variation on the theme of spectacular growth is that a few children race far ahead of the mainstream "pack" and are like small adults in vocabulary, grammar, conversation, and narrative skills, showing essentially adult skills by three years of age. For the top 2% of children who demonstrate this spectacular language growth rate, we have clues as to the emotionally rich and challenge-rich conversational opportunities that drive their growth.

One such child I came to know well because he, his parents, and his sister lived next door to me. The boy, Luke, by age three was reading complex plays and literature in both English and Hebrew, was a fluent conversationalist in both spoken languages, and was quite a marvelous chess player to boot! These remarkably early and varied skills were not triggered by instinct. Instead, they arose because of incredibly rich and supportive snowballing dynamic tricky mix episodes with his parents, and especially with his stay-at-home mother who is also an artist and writer.

## Creating Tricky Mixes that Make Bilingual Learning Spectacular

Paths to spectacular early bilingual growth begin with pre-language positive cycles of communication with loving partners. Children, if they are lucky, have social mixes that include in each target language three or four partners who love talking and playing one-on-one with the infant while the infant responds with eye gaze, laughter, smiling, and babbling. After these kinds of interactions in the first four months after a child is born, a strong foundation is in place for learning language structures within such communication across the ages of five-to-eighteen months. During this age span, children will make excellent progress in the phonologies of two languages as well as early vocabulary in the range of one hundred to six hundred words in each language by eighteen months of months. Then the dynamic tricky mix accelerator really takes off.

Now, between eighteen and forty-eight months, high rates of acquiring grammar and conversational skills are accompanied by continued high rates of vocabulary growth. These all work together to ensure spectacular progress toward fluent bilingualism when all of the following conditions keep contributing again and again to a powerful positive dynamic tricky mix. Moreover, some children encounter these rich mixes in four languages and by forty-eight months are fluent in all four!

- Conversations include highly positive emotions and fun
- Both child and adults have high expectations of enjoyable and loving interactions and of such positivity building across months
- Gains in language impact brain capacity so that the more the child has learned, the greater the brain's readiness for more learning
- Rich positive social interactions and neural capacity gains interact to promote accelerated knowledge of the real world—the world components that language then maps to
- Conversational strategies by adults with the child frame language challenges in ways that make the abstraction of new language structures easier
- A package of such adult strategies includes following the child's expressed meanings, leaving appropriate pauses that support deep processing, and the recasting of child utterances into similar but structurally challenging grammatical or narrative/discourse devices
- Some conversations every day carry in adult input well above the child's current level of language
- Cascading positive effects are produced in which all of the above conditions feed into improvements—beyond just maturation—of these cognitive processes: working memory, long-term memory, retrieval, perception, abstraction, processing speed, and executive function
- All of these gains in brain readiness then cascade and snowball positively across new rounds of conversation into ever higher rates of acquiring skill in lexicon, grammar, discourse, and narrative
- The child and their fluent adult/sibling partners sense the child's high rate of learning and keep raising their expectations

## When Evolutionary Changes Hit Turbo Speeds

We have seen in this chapter, as well as in earlier ones, many details about children's spectacular growth rates in language, art, and other domains. Now let's dive into the depths of evolutionary processes leading to changes in nonhuman animal species.

Only in the last two hundred years have enough scientists with enough fossil specimens and enough analytic tools in their hands arrived at convergent, widely agreed upon estimates of the typical rate of change in animal characteristics. This covers all kinds of animal features, including the first emergence of feathers, four-legged land animals, or animals like kangaroos and opossums with pouches. The typical rate is called one Darwin, and it is a 1% change in features each one million years—quite a slow rate, obviously.

Hmm. The next natural questions from a dynamic tricky mix perspective are: under what mixes of conditions are those change rates vastly improved, and what are the maximum rates ever observed? Prepare to be surprised.

It turns out that when a mix of environmental conditions shifts dramatically, so do the rates of evolutionary change in animals. This conclusion is supported by convergent, similar evidence from both naturalistic events and from experimental, causal events created by scientists.

As with the initially sad case from early in this book of a doggie who has lost the ability to climb simple stairs, the potential for change that exists is only seen when multiple relevant components of a dynamic mix are introduced.

So, how far above the usual one-percent-per-million-years rate can animal change rates rise? The answer is truly spectacular—60,000 times faster!

This rate has been documented in the Galapagos Islands—also a haunt of Darwin's—for multiple bird species by Jonathan Weiner. For example, cactus finches have co-evolved with cacti and rely upon cactus nectar for food just as the cacti rely upon the finches to fly from flower to flower and ensure pollen transfer and fertilization. Again, the cascade of environmental conditions that lead to measurable, spectacular rates of change in the beaks and other physical features of the birds can be summed up in a series which dynamically converges.

- Drought drastically reduces rainfall
- Less rainfall leads to fewer usual plant foods
- Alternative seeds & foods are difficult to eat given the bird's beak size and shape
- Less food consumed leads to less reproductive mating
- Less food consumed lowers survival rate for chicks

- Many birds in the species die. *Then*, only the birds with the most appropriate beaks and genes for this unusual dynamic mix of conditions will remain in the gene pool

Thus, beak sizes/shapes for survivors and their offspring have *shifted*, and evolutionary change has happened in just *one generation* and in just *four months*!

# Chapter 23

# COMPARING THE PRESENT "DYNAMIC TRICKY MIX" ACCOUNT OF COMPLEX CHANGE TO OTHER PERSPECTIVES

Space limitations compel us to bring in only a few authors and a few related perspectives. As you will see, there are areas of significant agreement as well as other areas of crucial contrasts with the present dynamic tricky mix framework.

## Tipping Points

Malcolm Gladwell gives an integrative account of many social and biological phenomena and some theoretical models trying to account for sudden change.

He argues that we are biased in our expectations toward thinking that changes will instead be gradual. "We are all, at heart, gradualist, our expectations set by the steady passage of time. But the world of the tipping point is a place where the unexpected becomes expected."

Gladwell goes on to stress his key metaphor, that of tipping points, where radical change is more than a possibility. It is—contrary to all expectations—a certainty.

"The name given to that one dramatic moment in an epidemic when everything can change all at once is the 'tipping point.'"

On this point, I agree that change can be rapid, as we have seen again and again already in this book. I also strongly agree that changes in our expectations toward or away from key beliefs are potent factors in how dynamic mixes unfold. But key differences with Gladwell reside in how dynamic tricky mix theory captures the dynamics of change varying from context to context and across diverse individuals and groups, and also in how many trajectories of change exist besides those that are sudden and pandemic-like.

So, let's look more closely at Gladwell's framework.

"The tipping point is the biography of an idea, and the idea is very simple. It is that the best way to understand the emergence of fashion trends, the ebb and flow of crime waves, or, for that matter, transformation of unknown books into bestsellers, or the rise of teenage smoking, or the phenomena of word of mouth, or any number of the other mysterious changes that mark everyday life is to think of them as epidemics. Ideas and products and messages and behaviors spread just like viruses do . . .

"Look at the world around you. It may seem like an immovable, implacable place. With the slightest push—in just the right place—it can be tipped."

As we will discover in the full range of discussions below, finding "just the right place" is not done easily. Instead, a more satisfying account of dynamic changes is one that better covers individual differences and complex converging factors and that explains what prior mixes/circumstances/events create likely tipping points versus more gradual actual changes.

Nevertheless, we want to be sure to give credit where it's due. There are some events where the tipping point metaphor captures much of what is occurring. And Malcolm Gladwell is lucid in some of his narratives about these events.

Among the "tipping" situations Gladwell covers are a contrast between particular "tipping actions," "nudges," or "tinkerings" within the high successes of the two long-running preschool TV programs *Sesame Street* and *Blue's Clues*.

"Every episode of *Blue's Clues* is constructed the same way. Steve, the host, presents the audience with a puzzle involving Blue, the animated dog. In one show, the challenge is to figure out Blue's favorite story . . . To help the audience unlock the puzzle, Blue leaves behind a series of clues, which are objects tattooed with one of her pawprints. In between the discovery of the clues, Steve plays a series of games—mini-puzzles—with the audience that are thematically related to the overall puzzle.

Whenever he asks a question, he pauses. But it is not a normal pause. It's a preschooler pause, several beats longer than any adult would ever wait for an answer. Eventually an unseen studio audience yells out a response. But the child at home is given the opportunity to shout out an answer of his own . . . If you watch *Blue's Clues* with a group of children, the success of this strategy is obvious. It's as if they're a group of die-hard Yankee fans at a baseball game."

I personally experienced the flavor of *Blue's Clues* by watching it with my own daughter, Leilani, when she was about four years old. She was indeed engrossed with the puzzles and with the doggy, Blue, and with the answers to the questions posed by Steve. Yes, indeed, she would jump up and shout out the answer with enthusiasm and accuracy.

Gladwell stresses that one key to a successful tipping toward high success both for *Sesame Street* and for *Blue's Clues* was to not change the basic content:

"Instead, they tipped the message by tinkering on the margin with the presentation of their ideas, by putting the muppet behind the *H-U-G*, by mixing Big Bird with the adults, by repeating episodes and skits more than once, by having Steve pause just a second longer than normal after he asks a question . . . The line between hostility and acceptance, in other words, between an epidemic that tips and one that does not, is sometimes a lot narrower than it seems"

## Infants Learning to Walk

At first glance, what we know about children's early walking might suggest a good fit with a "tipping point." After all, many of us have seen our own children go from crawling around one day to being upright and walking the next—of course, in a wobbly way initially. So, should we add this example to Gladwell's trophy list of tipping point phenomena, where change happens "all at once?"

Hmm. Maybe we shouldn't jump so fast to that conclusion. For infants launching into walking and for myriad other examples in business, biology, politics, medicine, and child development, we need more than anecdotal, common-sense observations. How "sudden" a change appears will depend upon the complexity and richness of observation. And it turns out for infant walking that we now have impressive detail about what leads up to the first walking a child shows. That complex data shatters most prior explanations of how walking comes about, both in general and for particular infants. Rather than a single, simple tipping point, multiple subcomponents of walking must move along fairly gradually and then achieve dynamic integration at multiple successive levels rather than through one giant leap forward. Satisfactory explanations so far lie in versions of Dynamic Systems Theory.

## General Dynamic Systems in Walking and More Domains

For Esther Thelen and Linda Smith, naturalistic, everyday recordings of infants as well as thoughtful experiments that use a treadmill and other special contexts have led them to place infant locomotor development within Dynamic Systems approaches adapted from biology, meteorology, economics, and other domains. They directly challenge other earlier views:

> "Both the maturationist and cognitivist views of local motor development assign a causal primacy to the mental structures that provided intentionality to the act of walking. That is, the shift from reflexive or more primitive movement to adaptive action is (held to be) at the level of conscious control of the leg movements.
>
> Here we challenge the implicit assumption of such single-cause models that a behavior such as walking "resides" in any single instantiation in the central nervous system (CNS) as either a neural or cognitive code. This assumption, that any behavior can be reduced to an essential that is represented in a privileged form within the organism, is widespread and seductive, but ultimately illusory."

Instead, Thelen and Smith cogently argue that a more complex way forward is needed at both the level of empirical observations and theoretical framing:

"To understand the developmental process through which multiple parts continuously and fluidly influence one another, we must study the dynamic organization of cognition as a complex system and empirically discover its points of stability, instability, organization, and reorganization."

## Dynamic Tricky Mix Theory for Walking and More

We have seen the dynamic tricky mix theory applied many times in this book. In the case of infant locomotion—crawling, standing with help, walking independently, running—it shows important contrasts with both the above views.

The "trickiness" of dynamic tricky mixes (DTMs) is in getting all the components of change lined up in real-time events. Infants are going to differ enormously in their temperaments, emotions, and expectations. Those factors then interact dynamically with the child's particular experiences of trying to move in different contexts. So, for example, a child of eleven months who seems fully "ready" in terms of brain maturation, muscle strength, vision, and general cognition may not begin walking until thirteen or fourteen months of age. This may happen if the child at eleven to twelve months is a low risk-taker, if they are easily contented with their current crawling ways of locomoting, and if they seldom see other children of ages eleven-to-twenty-four-months whose walking movements might be analyzed and who might inspire efforts at new movement strategies.

Accordingly, far more than the tipping point account, DTMs spell out the complex particulars that lie behind how a particular infant actually transitions to independent walking and then to running. And in comparison to the Thelen and Smith account within Dynamic Systems, the tricky mix approach incorporates not only their subcomponents such as perception, physiological awareness of bodily positions and states, attention, motor plans, memory, and anticipation, but also elucidates how further personal and contextual factors pattern together to support developmental advances. In consequence, it is straightforward in DTMs to expect a very wide set of trajectories for walking progress across different infants. Some will show very gradual advances overall, but also some short periods of very rapid tipping-point-like change. Others will show periods of seeming advance followed by oscillation/mixing between more sophisticated and less sophisticated walking movement patterns.

## Causal Experiments

Powerful insights into children's developmental processes as well as into change processes in all kinds of domains often emerge from casual experiments. In the case of infants moving around their environment, there are highly interesting studies. These show that when non-walking infants are given a controlled period of time moving and exploring around rooms in infant "scooters" before the infants are walking, a causal positive effect is generated. That is, infants given a "boost" of such practice are experiencing not just extra movement episodes but extra time relating their self-propelled movements to changes in the perceptual flow they encounter. In consequence, they show more rapid advance in perceptual sophistication compared to matched children who get no time in the scooters.

A strong conclusion can be drawn: the infants who got a mix of new scooter-based experiences were already prepared to benefit in terms of perceptual and motor skills. Their gains establish that no further maturation of the brain and no further general cognitive skills were required as a foundation for both perceptual and walking gains. Within our theoretical framework, we can say that their readiness for walking entered a broader dynamic tricky mix which triggered—not instantly, but over a few weeks—a cascade of small gains dynamically eventuating in walking.

Thus, we see a familiar pattern. For infant/toddler walking, old dogs with limited current movements, or children *stuck* in trying to acquire literacy, when certain new dynamic tricky mixes are introduced, they causally produce breakthrough achievements. Moreover, for all these instances, the breakthroughs demonstrate an already present set of readiness conditions which prior circumstances had failed to exploit toward needed advances.

## General Coping Styles, "Grit" and Related Concepts

When a person, group, or organization's long-term successes are considered, we may also see how other theoretical frameworks compare to the dynamic tricky mix accounts.

Consider the work of Angela Duckworth. Through a series of studies and interviews, she strongly stresses that the character trait of "grit" rather than inherent genius or other character traits is at the core of most long-term high achievement. The following short excerpt from *Grit: The Power of Passion and Perseverance* captures her central ideas:

> "To be gritty is to keep putting one foot in front of the other. To be gritty is to hold fast to an interesting and purposeful goal. To be gritty is to invest, day after week after year, in challenging practice. To be gritty is to fall down seven times and rise eight.
>
> "Grit holds special significance for the achievement of excellence. This is true whether the endeavor in question is physical, mental, entrepreneurial, civic, or artistic. When you look at the best of the best across domains, the combination of passion and perseverance sustained over the long-term is the common denominator."

From a dynamic tricky mix angle, these statements seem interesting but overly simplistic. And Duckworth does, at times, more modestly place grit in perspective.

"In the end, the plurality of character operates against any one virtue being uniquely important."

As just one example of a different body of research which also stresses multiple converging conditions in long-term achievement and in long-term mental and physical health, there is the longitudinal research of Valliant. A bit later we address Vaillant's work in some detail. But for the moment, in regard to Duckworth's conclusions, note that among the important strands of style and character stressed by Vaillant are a keen sense of humor, the tendency to play as well as work in a balanced way, and limited reliance on negative coping strategies such as projection of one's own anger onto others.

Moreover, so far Duckworth has no systematic account of how passion and perseverance are maintained over even short periods, let alone across decades of working toward a favored goal.

In tricky mixes, the dynamics of many converging conditions determine whether an individual or group is *stuck* in development, is moving slowly forward or backward, or is racing ahead with vigor and passion. Across this book's chapters, we see many patterns where causal changes in progress have been stimulated. And we see that snowballs toward more and more positive conditions often arise so that, for the same amount of time and effort, an individual has an excellent chance of sustaining high rates of achievement over long periods of time.

We suspect that, for the fortunate persons who experience highly positive snowball trajectories, those snowballs come in a wide variety of flavors. Again, we should expect no simple answers based on theory and research to date. Instead, we should study and seek to understand remarkably wide particulars in what personal characteristics; what patterns of interactions with potentially helpful friends, mentors, coaches, and others; and what mixes of life-enriching play and goal-centered efforts and strategies lead to long-term achievements and satisfactions.

## Daniel Pink's Concept of Drive

There is much overlap of the central concepts and researches we have just now reviewed above with ideas from Daniel Pink and also with my own concepts and examples concerning dynamic tricky mixes. Take, for example, the importance of individual autonomy and choice. Depending upon the other factors that are present and interacting, I agree that choice and autonomy are influential. In fact, our lab has shown through a favored dynamic tricky mix strategy of causal experiments that increased learner choices in literacy and narrative skill teaching raise engagement and the pace of learning. Mark Lepper and his colleagues at Stanford have done related causal experiments also showing powerful effects—more choices in math software exploration, for example, resulted in higher engagement and better learning by children.

Pink focuses especially on what business organizations could do—in our own theoretical terms—to re-mix the efforts they make to create plans/settings which dramatically improve both performance and employee and leader satisfactions. As we do with dynamic tricky mixes, he seeks to bypass simplistic goals and simplistic attempts to drive performance through usual concrete incentives. Instead, he argues that leaders and employees can collaborate to move everyone in their organization toward engagement of three instinctual, or nearly instinctual, "drives." First, autonomy—a need to give direction to our own lives and actions. Second, the need for mastery (please note the heavy overlap with what we have stressed in covering "self-efficacy" notions by David Kelley)—doing things well and improving in our skills, performance, and confidence. And third, the need to connect much of what we do to larger, socially valued purposes.

Bringing all of these together again and again is indeed a "tricky" balancing act. Pink provides intriguing case examples of business and education projects which have achieved the intended mix of motivational complexities and have achieved breakthrough performances.

Pink explores the new mixes that the software company Atlassian has created for its employees based upon a brainstormed idea of the company leader, Mike Cannon-Brookes. Two key components are to set aside twenty-four hours some weeks for code writers, software engineers, and others to work on any problem they like—as long as this problem is outside the box of their usual goals and projects. The fuller mix of conditions includes the freeing of multiple constraints: each person chooses not only their problem focus but also when they do the work and the team of others they collaborate with.

We have emphasized that for powerful tricky mixes to be found and sustained, organizations should achieve regular and valid monitoring of both dynamic processes and emerging wanted/unwanted outcomes. At Atlassian these both have worked out well, as summarized by Pink:

> "At 2:00 p.m. on a Thursday, the day begins. Engineers, including Cannon-Brookes himself, crush out new code or an elegant hack—anyway they want, with anyone they want. Many work through the night. Then, at 4:00 p.m. on Friday, they show the results to the rest of the company in a wild and woolly all-hands meeting stocked with ample quantities of cold beer and chocolate cake. Atlassian calls these twenty-four-hour bursts of freedom and creativity "FedEx Days"—because people have to deliver something overnight. And deliver Atlassians have . . . As one engineer observes: 'Some of the coolest stuff we have in our product today has come from FedEx days.'"

Going beyond these kinds of examples, in dynamic tricky mix theory we see that most breakthroughs have come about through bringing into convergence dynamically even more motivational factors than Pink highlights. Complexities abound! The joys of discovery, surprise, and humor mix together with multiple flavors of connecting to other complex human beings. More than that, over time, complex pathways to success emerge wherein part of the crucial tricky mix has been continual and open-minded monitoring of processes, adjustment of mixes on the basis of process-monitoring and experimental well-designed contrasts, and also an ego-free willingness to stop doing useless or destructive actions which had seemed highly promising until monitoring proved otherwise.

## Longitudinal Outcomes Between About Age Twenty-Two and Fifty-Five, Then Further Examined to About Age Seventy-Five

Time lags between what we do today in our mixes and what emerges as later outcomes in subsequent days, weeks, and years are complications in our determining what is working well and what is not, let alone what lies behind some breakthroughs. Longitudinal studies of individuals are one valuable window into such delayed dynamic effects.

As a thought experiment, ask yourself: What would be some of the tricky mix patterns for a person in their twenties which would be predictive of desired positive outcomes for the person by about age fifty-five?

Longitudinal research by George Vaillant and his colleagues at Harvard have yielded some tantalizing findings on this topic.

It turns out that a tricky mix "cluster" of characteristics for persons in their twenties does tend to predict a "cluster" of life successes over time, including all of these:

- More satisfying relationships
- More career successes
- Positive mental and physical health levels
- More income
- More joyful time in vacations & hobbies
- Deeper community involvement
- Successful aging

Across thirty years or more, each person, of course, will have a great many episodes which contribute to their career and relationship paths or other outcomes. The positive early "cluster" of characteristics for some folks, therefore, should be considered a kind of foundational tricky mix which again and again helps form new particular contextual tricky mixes that also incorporate other factors into each relevant episode. Positive cascades and snowballs of good processes, good decisions, and good persistence thus become more probable, even if not certain.

OK. Now we reveal from Vaillant's research the makeup in their twenties of those whose longitudinal life outcomes are positive across most or all domains of life. From interviews, tests, and more, the tricky mix most commonly seen looks like this in terms of mental/emotional coping styles:

- Humor is used to help with coping
- Provoking circumstances often trigger delayed rather than rash responses and active reflection before choosing a promising response
- Consultation or "referencing" with others feeds into most decision-making and into adjustments of plans
- High warmth in relationships began as children with parents and then fed into adult intimacy
- Relationships are valued and receive investment
- Work is balanced by regular vacations and fun
- Negatively projecting one's own feelings onto others is usually avoided
- Extreme fantasy is not impeding attention to realistic possibilities

*Then,* this definitely tricky mix of positive coping mechanisms will likely feed repeatedly over the years into more and more successful experiences and deepening emotional engagements. Then, the positive snowballs are likely to keep expanding as increased self-efficacy influences high curiosity and exploration as well as expanded social networks. These also enter into multiple new tricky mixes for work breakthroughs and intimacy breakthroughs. Intimacy begets more intimacy and provides a foundational security that is part of self-concept and which is a catalyst for "flowing" or "soaring" levels of dynamic mixes during many events.

Strong, positive overall health as well as excellent mental health further contribute to many of these snowballs over the years.

## Embracing Failure with Two Failure Gurus

Many episodes across prior chapters emphasize the high value of failures. Being *stuck* month after month is one pattern of failure. Unless something changes in our awareness or our efforts, then there is the risk that current failures will just beget more and more failure.

So, how does failure feed instead into breakthrough successes? Here is a maxim that introduces two failure gurus and sums up much of what we have already discussed:

> "Discrepancy-inducing and re-mix-inducing failures are powerful keys to breakthroughs."
>
> —Keith Nelson's Failure Maxim

Jack Matson's experiences with an unwanted disruption were described briefly in the chapter on dynamic puzzles. Now we will fill in some of what really happened in the decades following his frightful strike by lightning on a tennis court.

First, of the highest relevance to the present chapter, Jack developed and made influential the concept of "intelligent fast failure." In his innovative teaching at Penn State University, he has inspired students with his course, Failure 101. Jack has authored multiple influential books as well, including *Innovate or Die! A Personal Perspective on the Art of Innovation*.

Sometimes, re-mixing in our lives is irrevocably triggered by a sudden, unexpected event. So it was for Jack following his being severely hit by lightning. He no longer felt like himself, he could not remember important slices of his life, and he began to have depressive symptoms and suicidal thoughts. He was failing to cope well but had the awareness to seek help through psychotherapy. What he then experienced led to a familiar phenomenon we stressed earlier—that complex, positive changes in our self-concept will have widespread positive impacts on our decisions, our actions, and the mixes that we create, all of which shape our major projects and relationships.

In Jack's own words, here is how his therapist framed the work ahead from the very first session:

"He explained memory loss was especially critical because it defined who I was as a human being. The unrecoverable memory losses would change me as a person and would need be replaced by new memories. I would then become a blend of who I was and who I would become."

Jack's therapist led Jack to explore his own creativity, both through the therapy interactions and through totally new exploratory activities. As just one example, Jack for the first time tried to capture his experience through poetry. A series of new experiences further led Jack for the first time to an intense awareness of the fragility of our planet and its environment and a fresh commitment to weaving supportive, catalytic actions into his own professional agendas as a chemical engineer.

His new career pathways and projects as regards environmental issues has frequently led Jack and his associates into legal confrontations with giant corporations. We will frame these kinds of events within our dynamic tricky mix theory. Remember that the same underlying processes can support negative and nefarious snowballs as well as highly constructive positive snowballs.

When rogue corporations have for years grabbed high profits through a convergent and powerful mix of disregard for the impact of chemical toxins on humans and other species, lies, distortion of scientific and medical research findings, threats against critics, lavish spending on false media claims of corporate good deeds, and lavish expenditures for legal teams to try to squash any compensations to victims of their reckless assault on the environment, it requires some truly new bold, positive tricky mixes to yield any environmental justice outcomes.

Yet, despite the inherent trickiness involved, as an expert chemical witness in major legal proceedings, Jack has been again and again an essential part of legal victories for victims and for preventing others from becoming victims in the future. One such high-profile case has been that of Erin Brockovich as dramatized in the movie by the same name and by thorough investigative journalism. The Pacific Gas and Electric Company was proven to have severely polluted ground water resulting in public consumption of polluted, toxic water and, moreover, to have lied and covered up their misdeeds for many years. Jack is now professor emeritus from Penn State, but with his firm is still actively continuing this important work for environmental safety and justice.

After his recovery from lightning and the temporary disruptions to his self-integrity, Jack established a key guiding purpose in his life that decades later is still playing out in new mixes, new projects, and new relief against greed and reckless behavior.

Jack's ideas about intelligent fast failure (IFF) have also been playing out for decades in his own research and in his multiple creations of new collaborative institutes, teams, and projects primarily focused on innovation and entrepreneurs in engineering and business/corporate contexts. One beauty of IFF is that it carries within it a seeming tension. On the one hand, it provides a steady, unwavering attention to failures—there is a highly conscious, planful approach of embracing failures when they come along and seeking to analyze and learn from them; to do some "intelligent" re-mixing based on the carefully observed particulars of a failure. Yet, on the other hand, there is also an attitude of wild, bold, creative experiments and explorations and free-ranging generation of many ideas, of bringing highly diverse people and methods and ideas together to see what will emerge. The best of those kinds of explorations will be less carefully planned, often bordering on seeming craziness, with emerging consequences that were not anticipated. Out of such explorations there are excellent chances of new insights, genuine discoveries, and clear-cut breakthroughs along with, inevitably, some fresh failures to learn from.

Matson himself captures how his enthusiasm and overconfidence—a mix for stimulating creativity among undergraduates in a new class—fed into the wrong sort of dynamic pattern, into a negative snowball:

"[My approach] failed because of my dogmatic adherence to the short course syllabus. As such, I had unknowingly perfected "slow stupid failure" because I did not learn from my mistakes and continued to repeat the same blunders by not listening to the students."

There are so many adventures along the lines of intelligent fast failure which, instead, turned out very well indeed. One important example is the very tricky route he took to overcome opposition to the creation of a new institute at Penn State: The Leonhard Center for the Enhancement of Engineering Education, which Jack led for a good stretch. During that passionate stretch, he stimulated remarkable advances in entrepreneurship and the innovative attitudes and skill "tool kits" of engineering faculty as well as students. With colleagues, he also created a new university undergraduate minor entitled "Leadership" about innovation, creativity, and experiential learning in the service of educating engineering students to be better thinkers and innovators, whether in leadership roles or not.

Another direction of collaborative innovation has been his founding and continued influence on a new company, Envinity, dedicated to state-of-the-art choices in new and existing buildings.

## "Failure: Why Science s So Successful"

The book by this title is authored by Stuart Firestein. He argues that for all scientific endeavors it is essential to actively anticipate and embrace failures. But, by contrast, he argues that far too often science research is hampered by a too-simplistic, too-conservative bias to do "safe" experiments where the outcomes have been accurately predicted but without any generation of new, significant breakthroughs or depth of understanding.

He quotes the famed, highly creative scientist Enrico Fermi: "If your experiments succeed in proving the hypothesis, you have made a measurement; if they fail to prove the hypothesis, you have made a discovery."

His recommendations for improving science research are very akin to what we have seen in Matson's intelligent fast failure. By funding and hiring scientists who embrace high creativity, risk-taking, and abilities to learn from multiple failures, we can see that the directions he urges are highly likely to facilitate new and effective dynamic tricky mixes for scientific endeavors and an increased rate of achieving breakthroughs.

For example, he comments on a familiar scenario in which a faculty group has just witnessed a key talk by a job applicant with five years of planned research laid out and now must decide whether that person will likely be a highly original and productive professional in the future.

"The right question to ask is . . . what is the percentage of this plan that is likely to fail? It should be more than half—way more than half . . . because otherwise it is too simplistic, not adventurous enough . . . What kind of science would science be if it could make reliable predictions about stuff five years out? Science is about what we don't know yet and how we're going to get to know it"

In the same vein, Firestein laments frequent too-simplistic "bandwagon" stampedes to repeat a high-profile study without enough variations, creativity, and rethinking for the study repetitions to be worth the time, money, and hopes invested.

## Further Comments on the Views from Dynamic Tricky Mix Theory

Embracing failures and doing sufficient monitoring and analyses to learn from failures fits extremely well within dynamic tricky mix theory.

To a greater extent than other views, we strongly argue that the same basic Dynamic Systems underlie moderate rates of advances, much faster "flowing" or "soaring" rates, projects/goals that are *stuck* rather than moving, and projects/goals which are going downhill at varying rates. The different ongoing patterns all depend upon variations in the factors contributing to a particular project at a particular time, the intensities of each factor, and the strength of dynamic convergence among the factors.

As discussed especially in Chapter 11, quantum shifts, raising intensity levels on multiple factors, feeds into remarkable multiplicative, convergent advances in rates of progress toward goals—not just "motivation" levels, but a complex range of conditions which feed into high momentum *overall*. In consequence, rate differences on the order of 60,000 to 1 have been documented! So please keep complexity and intensity in mind in your projects and shoot for the moon!

In order to shift an individual or group project toward more favorable processes and outcomes, it will be essential, given the complexities of the dynamics which apply, to invest considerable time and resources into monitoring frequently and then adjusting mixes accordingly. Failures of many sorts will, of course, receive this monitoring, as will successes. Whenever possible, trying out many experimental mix variations simultaneously—or at the very least, without much time lag between the experiments—will be a powerful overall strategy.

# Chapter 24

# ARE THERE INSTINCTS FOR THE MOST COMPLEX HUMAN PERFORMANCES AND ABILITIES?

"To say language is not innate is to say that there is no difference between my granddaughter, a rock, and a rabbit. In other words, if you take a rock, a rabbit, and my granddaughter and put them in a community where people are talking English, they'll all learn English."

—Noam Chomsky

Here we must note that, despite Chomsky, in America there was a short-lived fad of adopting pet rocks. Many crazy giftings of these rocks were accompanied by notes praising what minimal care these "pets" required compared with other pets. To date, Chomsky would be pleased to note that no cases have been reported of the rocks learning English or any other language.

But now it's time to look more closely at ideas and data concerning human instincts. Language and music are among the most exciting and complex domains in every human culture. They are both highly complex in terms of the structures that are coordinated for effective communication. And they are both truly complex in terms of the speed at which individuals perform and mesh the timing of their performance with other humans. Later we will consider other domains of complex human performance as well. For all such domains, an enduring question for science has been the source of complex performances in humans that are beyond any seen in other species. It has been tempting to multiple scholars to yield to a simplistic biological explanation—instincts which are built into the human species are claimed to provide a full explanation.

## Recap on Too-Simple Language Assumptions

Aristotle made the far-too-simple claim that language is acquired because there is a language instinct. In modern times, Chomsky and Pinker have made precisely the same claim but clothed it in various descriptions of linguistics and language differences and various other rhetorical means of trying to make the claim seem plausible. However, in their writings they still were not escaping the fundamental circularity that trapped Aristotle. That is, the primary claim that "proof" of the language instinct is that most children acquire considerably complex language and the reason they are able to acquire that language is because of the language instinct. Clearly, this is just a complete circle of reasoning.

As brain imaging techniques over the last fifteen years became increasingly powerful, it was widely believed among nativist theorists that a picture would emerge of a powerful, small language organ in the brain sitting in plain sight in the imaging records. This has proven to be a chimera.

Instead, the modern, evidence-based persuasive accounts of how a child acquires one or more languages in the first six years of life is remarkably more interesting than the instinctual attempts at explanation. Under the right complex and dynamic conditions, most children are able to proceed smoothly in the acquisition of one, two, three, or four languages by about six years old. Obviously, when four languages are acquired in this time frame, it is impossible for the child to be in constant contact with excellent learning conditions for each of these languages. All the child requires is a modest amount of truly well-mixed tricky, dynamic, engaging conditions. The child requires enjoyable, socially engaging interactions in which specific kinds of discourse exchanges are embedded.

Under highly favorable positive dynamic tricky mix conditions, the child will encounter examples of language complexity used by adults and older children and will abstract new levels of language structure week by week because conditions beyond mere exposure to fluency are present. In a typical rich conversational exchange, the child's message will immediately receive a relevant adult reply. In the reply, the fluent speaker will not just display complexity beyond the child's level, embedding needed challenges into the flowing discourse. Instead, the child's partner also will maintain positive emotional and social conditions and high attention from the child. In consequence of this rich child-specific conversational tricky mix, the child's powerful learning mechanisms will move like a laser to perform an analysis/abstraction of the contrast between their own language output and the output of the fluent speaker.

Thus, the powerful use by the child of their cognitive capacities is dramatically advanced when there is low anxiety, high social engagement, positive emotional states, and high expectations that the discourse will be interesting, fun, and enjoyable. The dynamic, powerful cognitive abstraction at the heart of language acquisition is richly driven by the accompanying attentional, social, emotional, and engagement conditions. They all work together convergently to create what seems like a "miracle" of language progress by children who are still, at ages one to three years, also in the process of developing increasingly complex cognitive and social skills.

One aspect of the cognitive abstraction processes at play even at two and three years of age is the ability to abstract common patterns from examples which vary greatly in some characteristics. Not all lobsters, dogs, or "Fiffins" (special toy invented for vocabulary experiments) look the same in most details, yet young children readily abstract/extract the underlying common features of each concept. Ditto for grammatical structures such as past tense or future tense.

Another aspect of the cognitive abstraction processes is that the child language learner extrapolates what they have already learned. As part of an earlier puzzle, we gave the example of a young girl saying, "Yestermorning I maked a tiger." This entirely original sentence makes perfect sense, and "yestermorning" is an actively created new word extrapolated from already-understood words like "yesterday." Similarly, a three-year-old boy who knows that he should not take toys from a baby extends that knowledge. He says, when mom introduces very small "baby" peas for dinner, that "We can't eat baby peas!" Funny for sure, and original for sure. What he means is if those are the baby's peas, then it would be naughty to eat them.

At this point, dear reader, the current chapter has provided a clear "solution" to Dynamic Puzzle #10 of Chapter 5. Modern research and theory development delivers a highly plausible alternative to views that language in humans is possible only because a "language instinct" exists. Children usually move along well in their acquisition of language in the first six years of life *because* in meaningful social interactions they bring their powerful active minds into encounters with highly favorable positive dynamic tricky mix conditions.

Below we will review one particularly informative causative experimental study which showcases the complexity of interacting cognitive processes and the challenges and supports within child/adult interactions. The discussion of the research findings will carry us back to the biological/evolutionary processes already raised in Chapter 15 concerning fish, giraffes, leopards, and other creatures and their forms.

## The Minds That Help Build Themselves

In the wide, wide world of animal species, some remarkable evolutionary successes rest upon highly specific neural networks built into an animal's young brain. For example, stickleback fish who respond with aggression to a red dot on another fish. And herring gull chicks who respond with hunger and pecking to a red dot on an adult's beak. Or Emperor Penguin males who employ a special egg-ready warm, lined pouch to protect and incubate a single penguin egg in the cold, windy, harsh Antarctic climate.

For the young of our species, it's a whole different story. For infants and toddlers and their remarkable progress toward complex communication and other behaviors crucial to our species, the evolutionary secret to key successes is to *not* building the brain with highly specific ready-to-fire circuits.

So, we may rephrase this conclusion. The breakthrough in evolutionary brain design is to prepare the young human brain for learning complex systems of communication by building in highly flexible learning and performance capacity rather than fixed, prepared patterns specific to communication. We will elaborate below on evidence supporting this idea. Here, note that other scholars have addressed overlapping notions about "emergentist" acquisition processes—addressing both biological makeup and actual social interactive patterns—for language and other complex human systems.

## Underlying Cognitive Processes are Dynamically Shifted in Their Deployment to Support Different Domains of Communication

Both in learning and performance there are multiple cognitive processes that are essential, and the patterns of their dynamic ongoing activation/deployment must shift depending on modality. In the Resources section at the end of the book there are good illustrations of these variations by modality. Recent research has emerged that specifies how strongly each of the component processes are involved for each of these contrasting modalities: listening to music, playing music, comprehending or telling a narrative, accessing vocabulary, learning a second language, and producing appropriate syntax/grammar.

As for the cognitive processes of relevance, here is a partial list: abstracting/learning new patterns, working memory for extremely recent events and goals, long-term vast memory resources, "executive" planning and strategies, integration of emotion, attention, and monitoring. One surprise has come from relatively recent complex analyses which assess the *relative* contributions of these different dynamic mix components—working memory is usually one of the weaker contributors. A sample of our relevant studies across different language domains and different child and adult groups is provided in the Resources.

## Children's Vocabulary Acquisition Processes as Revealed Through Causal Intervention within the Framework of Dynamic Tricky Mixes

Marnie Arkenberg and I have provided a full discussion on dynamic tricky mix theory in regard to powerful evidence of high rates of language learning possible when the dynamics of the learning situation are changed. In her study in my lab, two groups of four-year-olds were matched in vocabulary knowledge, awareness of print concepts, working memory, phonological awareness, attention, and ability to generalize. One group of children participated in a twelve-week word-learning intervention where they were exposed to four hundred fifty unfamiliar animal names. Note that this a *large* set of challenges. In accordance with dynamic tricky mix theory, children were exposed to these new words under highly engaging, child-directed sessions in which the investigator followed the lead of the child by allowing the child to dictate the next animal to be learned. Children also chose which of many playful activities to start next. This scenario translated into a dynamic mix we would expect to create high rates of learning.

Results showed that when children were taught new lexical items during these intensive, rich, positive, playful, and highly engaging episodes, they showed exceptionally high gains of up to twenty words per hour. This is a remarkable learning rate! By way of contrast, estimates of fairly typical naturalistic rates for four-year-olds are much lower, at between two and twenty words per *day*. That such a high rate of acquisition can occur suggests children certainly have the capability to rapidly—learn many new words when the dynamic tricky mix conditions for learning are especially favorable.

In addition, it appears that multiple cognitive gains contributed to these findings. Because we assessed cognitive ability in the two groups of children prior to the intervention, halfway through the intervention and at the end of the intervention, we were able to compare changes in working memory, phonological awareness, attention, and abstraction/generalization abilities for children who did and did not participate in the intervention. While children in both groups were well matched on these cognitive abilities prior to the intervention, we saw significant cognitive gains across time only for children in the intervention group.

Moreover, because the boost in their cognitive abilities had already begun one month into the study, that cognitive boost dynamically contributed to further, enhanced language learning rates across the final three months of the vocabulary intervention sessions. To date, this experimental causal study provides the clearest evidence in the field of how enriched and highly challenging dynamic social, interactive mixes causally facilitate both enhanced learning rates and enhanced neural capacities for future learning. This is a beautiful example of complex, interrelated changes comprising extremely positive dynamic snowballs.

## Evo-Devo Comparison

The dynamic processes in child development just covered above echo other related kinds of dynamic processes, both in evolution and in the embryological development processes leading to the emergence of a newborn infant animal—whether human or zebra or dragonfly or chimpanzee or any other animal species. These latter processes have been dubbed "evo-devo" for "evolutionary developmental biology." Sean Carroll, the distinguished geneticist, describes them as follows:

> "The development of form depends upon the turning on and off of genes at different times and places in the course of development. Differences in form arise from evolutionary changes in where and when genes are used, especially those genes that affect the number, shape, or size of a structure...
>
> "Regulatory DNA is organized into fantastic little devices that integrate information about position in the embryo and the time of development. The output of these devices is ultimately transformed into pieces of anatomy that make up animal forms. This regulatory DNA contains the instructions for building anatomy, and evolutionary changes within this regulatory DNA lead to the diversity of form."

In our capsule summary of the vocabulary-enhancement intervention above, we concluded that the very neural processes and cognitive components which support learning were themselves enhanced because so much rapid learning occurred. In all likelihood, this rested upon the new neural networks established for new vocabulary items then triggering gene regulatory devices which helped provide excellent neurotransmitter conditions for further learning. For example, if the genes necessary for production of neurotransmitters were turned on and off in a more optimal pattern for the children with the big vocabulary gains, then this could help set off a cascade of further positive effects—faster learning, enhanced working memory and attentional and abstraction processes, and more and more optimal availability and balance of neurotransmitters. All of this fits perfectly within modern evo-devo research and theorizing.

## Chapter 25

# BUTTERFLIES AND DRAGONFLIES

A butterfly moving from flower to flower in a rain forest in Honduras will no doubt in the next few moments—while you read this passage—be subject to many dynamic, complexly mixed local conditions. Beyond that, Dynamic Systems Theories and Chaos Theories argue that what this butterfly does in those few moments will sometimes be a crucial part of a cascade of events that, over time, lead to a dramatic climactic event such as a typhoon. Air currents influence the butterfly, and then the butterfly's movements minutely influence new wind patterns in such a way that new wind, temperature, rain, and ocean wave patterns unfold toward a typhoon versus toward a mass of warm, mild, precipitation-free air.

For humans aspiring to be mindful and to steer the biosphere in positive directions whenever possible, it is important to take seriously the notion that planful current actions sometimes will dynamically lead to a cascade of effects that are far-reaching and positive.

Douglas Tallamy, an entomologist, through his research and writing explores the surprisingly strong effects of landscape and gardening decisions by homeowners on the diversity of insect and bird species. For the most part, decisions about what shrubs, trees, lawns, and flowers go into our backyards have focused on appearance only and have ignored possible dynamic effects on any animal species. These decisions have introduced a very large number of non-native species into North American landscapes. So what? Well, Dr. Tallamy's findings are that suburban landscapes dominated by regularly mowed, grassy lawns and non-native ornamental species are a disaster for biodiversity.

Dynamically, the negative cascades are that non-native plants host very few insect species and, in turn, few small bird species, frogs, and toads. In turn, these declining species lead to fewer and fewer hawks and mammals that depend upon the birds and amphibians as food. Dynamically, moving away from native species of plants reduces the diversity of nearly all kinds of animals in the environment. And these effects are often huge. For example, in Delaware backyards, native white pines have been shown to host more than two hundred times the insect species as the most common non-native trees. As another example, consider that in many ecological contexts a single "alien" invasive species can crowd out almost all other plants. Here is Tallamy on invasive olive trees:

> "A field rich in goldenrod, Joe-Pye weed, boneset milkweed, black-eyed Susan, and dozens of other productive perennials supplies copious amounts of insect biomass for birds to rear their young. After it has been invaded by autumn olive or Russian olive, that same field is nearly sterile."

For biodiversity, we once again see that the same dynamic mix processes—given different particular mixes—yield incredibly varied outcomes from biodiversity dropping in suburbs and radically snowballing toward sterile areas to flourishing, positive snowballing native species diversity both for plants and animals. Butterflies, as part of these ecosystems, rise and fall in concert with the overall diversity of an ecosystem.

## Dynamic Tricky Mixes for Monarchs—Negative or Positive

A huge decline in magnificent orange and black monarch butterflies in recent years was at first a puzzling event. But then studies revealed just how much the monarchs are impacted by new large-scale, toxic farming methods. All across the US Midwest there has been a vast increase in Monsanto-inspired use of genetically modified corn and soybeans and the closely coupled promotion of vast amounts of the herbicide glyphosate (the active ingredient in Roundup) that leaves the new genetically modified corn and soybeans unharmed. Millions of acres have diverse plant species destroyed as a result, including the keystone, critical plant for monarchs—milkweed. Now monarchs migrating north from Mexico to mate in diverse US areas all the way into Canada must find the much rarer milkweed plants on the relatively small acreage that is spared from modern toxic farming. Without the milkweed, new monarchs are not generated because the only place female monarchs lay their eggs is on milkweed plants. Thus, in recent years the complex yearly cycle for monarchs is less often successful.

Humans are now in the position of saving the monarchs by creating re-mixes of conditions using the same dynamic principles and processes that are driving their numbers downward. My firsthand experiences with monarch ecology and possible re-mixes of local conditions in Pennsylvania are many. As I write here today at Eagle Spirits Farm, because of many years of encouraging milkweed, there are far more milkweed plants than before. These plants are mature and show many small, white eggs laid by female monarchs. And my wife, Kathy, and I daily make sightings of the monarchs both near milkweed and an array of diverse flowers which provide them food.

My students at Penn State University and I have also made related efforts. We raised over 2,000 milkweed plants in an available research greenhouse. These were then transplanted to fields all around our area. Fifth grade children and their teachers learned about monarch life cycles and assisted in the transplant operations. And to aid research by the international project Monarch Watch, we placed super-light numbered tags on the delicate wings of monarchs. Believe me, it was a thrill to see a September-born monarch with its fresh tag fly vigorously off from our Pennsylvania farm in a southwesterly direction. That flight plan might well lead—given just a few more fortunate weather and food conditions mixed in—to a winter stretch of life in Mexico.

As a society, we also could cooperate and move toward mass farming techniques that limit, rather than celebrate, herbicides and pesticides. Milkweed extent and diversity would then be promoted along with plant diversity of myriad sorts, including a vast array of plant species that support the overall ecosystems. And as the plant diversity expands when the farming methods evolve, so will the diversity of other insects, amphibians, songbirds, carnivores, and many other mammals.

To go back to the "butterfly effect" metaphor, a new and very satisfying set of effects could be generated by good planning and monitoring of farming. As monarch numbers expand by many millions along with their essential milkweed, there will be cascading positive dynamic effects over and over, generation after generation, on the richness and sustainability of plant and animal species. As these ecosystems are restored and enhanced, the opportunities will cascade upward for children and adults to enjoy more frequently not only Monarchs but also so many other components of the natural world. More wildflowers, more songbirds, more amphibians, more herons and eagles, more wild species of every sort.

Images below bear on the life cycle of monarch butterflies, the emergence of forms from adult butterfly to egg to larva to chrysalis, and the co-evolution of needed milkweed plants and the butterfly species in all its phases.

*Here at our farm, we have helped expand milkweed plantings. Milkweed co-evolved with monarchs and are essential to life cycles of the butterflies as sites for eggs and caterpillars.*

*Depending on what month they are born, monarchs in butterfly phase live for either about one month or about seven months. They eat nectar and help to cross-pollinate milkweed and other flowers.*

*The orange and black wings of a butterfly inside its chrysalis just before the chrysalis explodes open and the compressed, wet monarch emerges.*

## Dragonfly Discoveries

Herding sheep is quite easy, herding cattle is still fairly easy, herding cats is very difficult, and herding dragonflies is near impossible. Why? Well perhaps it's because they are nature's most amazing fliers. Biologists trying to catch and examine them find that they are "chasing little rockets," according to Kent McFarland of the Vermont Center for Ecostudies.

Among the fastest of flying insects, dragonflies are masters of the air around ponds and other water sites. Their remarkable physical structure and impressive aerial movements have inspired human engineers to create, among other devices, helicopters and drones.

Two sets of swiveling wings allow them to hover motionless, fly upside down or backwards, pivot direction at lightning speed, and also shift rates of movement suddenly. Further, huge, multifaceted eyes allow for nearly three-hundred-sixty-degree acute vision. Dragonflies' rare and tricky mix of combined characteristics—speed, agility, transparency from certain angles, aggressiveness, and superhero vision—make them breakthrough, superbly skilled predators. They are able to track, grab, and devour insect prey in midair, often sneaking up from below. As predators, they far outpace the efficiency of hawks or eagles; dragonflies successfully catch their targets more than 95% of the time. Their prey includes many insects that transmit disease or ruin picnics, including mosquitoes, gnats, horseflies, midges, flying ants, and termites.

Suspicions that dragonflies were taking their aerial display shows on the road—that is, migrating—have been around for a long time. Historical records show some awareness of such migrations more than two hundred years ago. Nevertheless, scientists until a few years ago were stymied in trying to capture the detailed migratory patterns of whichever dragonflies were migrating.

New mixes of materials in studying migration included better superglues, ultra-light physical tags, and miniature radio transmitter devices. The first round of tagged dragonfly research revealed a greater range of travel during migration than had been known. Then, much finer further detail emerged when continuous streams of data were collected from raindrop-sized transmitters placed on the dragonflies. We now know that the champion for long distance dragonfly migration is the globe skimmer with its travels across the Indian Ocean between India and Africa. They travel up to 10,000 miles in a cycle. Very rich detail on migratory locations and timing have also been revealed by similar transmitter studies for the beautiful green darner dragonfly. More creative tools employed for this species are considered immediately after this striking image.

*Green Darner Dragonfly*

Creativity often involves seeing what no one else can see. You may have personally demonstrated a version of such creativity in viewing and reacting to photographs laid out for you in earlier chapters.

Here's how an intriguing breakthrough emerged for dragonfly research. A new study team went to museum collections where the wings of green darners could be analyzed for certain tag molecules. These hydrogen molecules were strong clues to where a dragonfly was born. By comparing the results to where the dragonfly was found when collected, whether or not the dragonfly had migrated could be determined. Three different isotopes of hydrogen exist, and they vary according to water temperature and climate. This group of scientists calculated the proportion of southern-born dragonflies in the USA by examining the proportions of different isotopes of hydrogen. Before this, no one had seen the isotopes of hydrogen in the dragonfly wings, and no one had really seen the full pattern of migration of any species of dragonfly.

The breakthrough pattern revealed by this research is astonishing. Dragonflies born in the southern United States emerge from their aquatic nymph stage, then unfurl and dry their wings for an hour or two. Then they fly an average of four hundred miles north, with some traveling eight hundred miles, to their new homes along ponds. There, that first generation mates, lays eggs in wetlands and ponds, and dies—a total adult lifespan of less than two months. Once the nymphs of this second generation reach adulthood, they'll swarm south, mate, lay eggs, and die. Those eggs become the third generation, which typically stays put with no travel demands. These dragonflies just hang around and party in the sun, go to the beaches and such. This third generation enjoys all kinds of incredible social cycles and have a high old time. Gnat-munching parties are especially popular. For this third generation, their offspring will eventually zoom northward, following in the pathways of their great grandparents. Thus, the dynamic cycle begins anew.

## More Parallels Between Butterfly Ecology and Dragonfly Ecology

Migratory behavior is fairly common for bird species, but comparatively rare for dragonflies and butterflies. In all cases of spring migratory patterns, complex dynamics involve how individuals negotiate travel routes, how their timing fits in with the differential food, mating, nesting/egg-laying, shelter, and water resources available. In turn, if we as humans want to facilitate the resilience and abundance of diverse native species, whether dragonflies or butterflies or birds, our own action patterns need to respect and converge with dynamic patterns already in play in nature.

## Why Should Humans Pay Attention to Nature, and How Can We Expand our Playbook of Dynamic Re-Mixing Strategies?

Peering into the complexity of dynamic processes in nature, including evolving ecosystems across many millions of years preceding any evidence of human-like creatures, has high relevance for how we approach important aspects of modern life. These include economic, social, health, education, justice, governance, and entrepreneurship issues and decisions. If we reveal the "playbooks" of converging factors and events behind fascinating advances in the natural world, those playbooks can be creatively adapted when we aim for re-mixes of human conditions in almost any domain. More than that, by knowing the complex dynamics of particular places on this earth and the ecological patterns therein, we are laying the necessary foundation for creative human tricky mixes which can be put in place to preserve and expand the vitality of the ecosystems. Nature is prepared, as well, to provide rich gifts for humankind. If we succeed in research to understand nature's patterns, new discoveries will lead to gifts for us in the forms of medicines, novel materials for technology, agricultural innovations, and inspirations for the arts and sciences.

Given all the above considerations, there is urgency for new creative mixes of efforts to help biodiversity recover and flourish in ecologies throughout the world. Tallamy is very clear on this issue:

> "Studies by dozens of scientists all suggest that ecosystems with more species function with more efficiency, are better able to withstand disturbances, are more productive, and can repel alien invasions better than ecosystems with fewer species . . . Biodiversity is a national treasure that we have abused terribly, partly because we have not understood the consequences of doing so. Our understanding of such consequences is far from perfect, but we now know enough to behave responsibly toward the plants and animals on which we ourselves depend. We must manage our biodiversity just as we manage our water resources, our clean air, and our energy."

Here we should invoke the Keith Nelson corollary of the Warren Buffet axiom:

> "It is not necessary to do extraordinary things in order to achieve extraordinary results."
>
> —Warren Buffet

"To achieve extraordinary results, it is necessary to bring a mental model—although imperfect—of relevant, complex conditions in dynamic interplay and to make action proposals respecting those dynamics."

—Keith Nelson's Corollary

Once we come to know and respect the complexities of biodiversity and the rich contributions of native plants in ecosystems, there is some danger that we might feel immobilized by the complexity we see. Instead, we need to brainstorm high-leverage strategies and pivot vigorously toward practical actions that will help to start significant positive snowballs on their way. Tallamy gives one wonderful example along these lines:

"Increasing the percentage of native species in suburbia is a grassroots solution to the extinction crisis. To succeed, we do not need to invoke governmental action; we do not need to purchase large tracts of pristine habitat that no longer exist; we do not need to limit ourselves to sending money to national and international conservation organizations and hoping it will be used productively. We can each make a measurable difference almost immediately by planting a native nearby. As gardeners and stewards of our land, we have never been so empowered—and the ecological stakes have never been higher."

## Causal Experiments on Raising Awareness of Nature's Patterns and Our Opportunities to Be Good Stewards

Oftentimes it may seem that any form of new, direct encounters with nature will have positive effects on the child or adult who participates. Perhaps. But how do we know the effects?

Again, we must note that activities will be more promising if they take into account the complexities of nature's patterns and the complexities of how children and adults may achieve significant advances. Research that actually creates dynamic tricky mix interventions within that framework and monitors their effects is seldom undertaken.

One key exception is the study conducted at Penn State University in 2017 by Alyssa San Jose and myself in which both children enrolled in a nature camp and their counselors showed significant advances as a result of the five-day program. They increased their knowledge of nature, their positive attitudes, their awareness of dynamic patterns, and their inclination to personally take actions to support nature.

New cycles of careful, monitored efforts in this field are needed in the USA and abroad to support improved effectiveness for whatever time and resources are devoted to enhancing nature and our lives through contact with nature.

Chapter 26

# PERFECT DYNAMIC STORM: HOW SIMPLICITIES RUIN CRIMINAL JUSTICE REFORM IN THE USA AND ABROAD

I n Criminal Justice systems in the USA most attempts at reform have been bitter failures. We will give a Dynamic Tricky Mix account of why that is case. This sets the stage for new approaches that hold high promise.

First, a recap of a few themes.

As in prior chapters, we will see here that for long stretches of time, too-simplistic packages of strategies are put in place with terrible results. Among the counterproductive approaches are "choose the best and forget the rest," "when you fervently believe your plan will work, don't spend resources on timely evaluations," and "stay snug in your silo—don't bring in outside-the-box thinkers and consultants."

## Prisons and Prison Programs in the USA

The record is obvious and shameful.

Compared to other modern, industrialized countries, the USA throws far more individuals into prison and does a far less effective job of improving the skills and later-life successes of those who have been in prison.

So, what lies behind the repeated failures of the USA in trying to create effective dynamic mixes for low criminality and for high potency of rehabilitation programs?

Blocking progress on these issues are a series of way-too-simple assumptions, conclusions, and strategies. In this regard, here is a partial list that will be useful when we try to capture the toxic dynamic mixes that have been set up in the US criminal justice system:

1. Privately run prisons provide both lower costs and better rehabilitation.

2. It is assumed that basic language skills in adolescents and adults are beyond change. This conclusion is not based upon evidence but rather on outdated notions that language learning abilities are "innate" and that language must be learned as a young child before the mind's flexibility snaps shut.

3. Accordingly, prisons use the strategy of not treating language deficiencies in prisoners despite clear evidence that huge language deficiencies are typical of prisoners in every country ever studied.

4. There are limited self-management opportunities in prison, stressing instead prisoner restrictions and control procedures.

5. Nothing is borrowed from Scandinavian countries with a far better criminal justice record than the USA because those countries are too different from the USA and too "socialist."

Remember that the same set of dynamics are at work regardless of whether an individual's rate of progress is positive and fast, just so-so, or rapidly changing in a negative direction.

Persistent negative Dynamic Tricky Mixes have been causing multiple negative effects not only in the short-term, but year after year along a person's life trajectory.

Many persons who end up in jail or prison got off on the wrong foot even before kindergarten, with language skills far behind their peers.

Then in school, for such children, time is spent almost entirely on activities that provide no decent opportunities for gaining the needed communication skills. Consequences often include poor academic performance, low self-esteem, and poor relationships with other children in grades one and two.

The same dynamic mixes overall tend to persist into later elementary grades and through high school, but with even less favorable conditions because the child develops negative reputations and expectations so that other children and teachers feed zero or low growth in skills. The child's own increasingly negative self-expectations and self-worth add fuel to the dynamic fires for low growth in cognitive and social skills and for increasingly negative attitudes. Peers are chosen as friends and companions who have similar characteristics, and deviant and illegal behavior frequently flow from all these dynamic conditions mixing together year after year. In sum, powerful negative and snowballing tricky mixes result in high criminality.

Once incarcerated, the contextual conditions are even worse than in schools. The likelihood of favorable dynamic mixes for growth in any of these skill and motivation areas is extremely low: oral communication, cooperation, literacy, problem-solving, specific job skills, self-regulation, confidence, and motivations for competence and social contributions.

Each time a released prisoner returns to incarceration, the dynamic mixes worsen further. No wonder that under US criminal justice there is such a poor record of rehabilitation of prisoners to a life course of constructive contributions to society and avoidance of illegal behaviors.

## A New Approach

A fundamental re-mix of conditions in jails and prisons is desperately needed. So, reader, here's your chance. Design your own new mix of conditions that will change a man's life course before you read on. Thank you!

Now, we do have a few ideas of our own on this topic.

The four most central strands of a re-mix we recommend are covered below:

1. Since children with severe delays in communication can reach main-stream levels when radically new conversational conditions are set up, as we have seen earlier in this book, borrow these proven procedures used in treating children to support needed strong language growth in prisoners.

   The dire communication status of male prisoners in the USA and abroad almost never (currently) results in language treatment procedures of any kind. This holds true despite the fact that in every country so-far studied, at least two-thirds of male prisoners have clinically significant language delays. For these prisoners, their low language levels severely hamper their learning ability and their social interaction in myriad situations.

2. Once deficient communication levels are overcome for prisoners, other learning opportunities will be re-mixed because the language-rehabilitated prisoner will bring better comprehension and expression into all contexts. Now they will understand more and learn more from educational classes as well as from all approaches to drug avoidance, self-regulation, constructive rather than criminal strategies, social relationships, and literacy.

3. Borrow extensively from Scandinavian rehabilitation programs for prisoners. Place responsibilities for self-management and planning upon prisoners, but support that with friendly and optimistic staff. The staff should expect (realistically, from prior successes) that by practicing constructive and planful social and cognitive behaviors in prison, individuals will be prepared to continue those behaviors in any new context outside of prison. All of the staff-prisoner interactions will dynamically benefit from the advances in communication skills that are fostered. In consequence, positive spirals and/or snowballs of better and better skills and better and better attitudes and confidence will be produced. Thus, rates of positive change for the prisoners will keep accelerating over time. "Flowing" or "soaring" progress levels will be common.

   Norway has a superior record in implementing these kinds of complex, restorative, truly rehabilitative tricky mix prison programs.

4. Dramatically alter the levels of challenge and the level of expectations for academic programs in prisons.

Very unfortunate, counterproductive, negatively snowballing dynamic cycles are typically set up once a man is convicted of crime and sentenced to a high security prison in the USA, as we have noted. The majority of incarcerated men are *stuck* in complex, dynamic conditions that block their rehabilitation in terms of any life-changing advances in communication, job skills, education, confidence, social skills, self-management skills, or networks of constructive contacts.

However, in the next section, we will look at a remarkable exception to the prison experience and life once released. Namely, we will see that in one program, two breakthroughs are achieved. First, the men complete a college degree. Second, after they are released, they nearly always avoid returning to prison. 96% of the time, these men live constructive lives and stay out of future criminal behavior and out of prisons.

## Bard Prison Initiative

The Bard Prison Initiative (BPI) extends the full breadth and depth of the Bard College liberal arts curriculum into six New York state prisons. The program's success in implementing innovative education programs for prisoners are among the best-documented. Its positive dynamic tricky mix has these converging components: live, in-class discussion with the professors; challenges that are high and well above what most educators have presented prisoners; exceptional commitment to the program by the students; positive feedback cycles to currently enrolled prisoners from graduates of the program now in constructive roles after release from prison; changes, as we have stressed as powerful advances in other contexts, toward healthier self-concepts and self-efficacy; high emotional engagement; support from fellow student prisoners; and realistic schedules allowing enough time for study and for mastery of educational goals.

College faculty teach small, seminar-style courses. In addition, the professors serve as academic advisors and offer workshops. Students enroll full-time in the same courses that they would on Bard's main campus. More than five hundred Bard College degrees have been earned by these incarcerated students.

More recently, other universities and colleges are also implementing efforts that support prisoners earning degrees while they are in prison. These include the John Jay College of Criminal Justice, Cornell University, Northwestern University, and California State University.

The broad successes outside of prison by the Bard graduates, we argue, depend upon the complete tricky mix we have summarized above—not just upon having a college degree. The fact that 96% of graduates released into society do not commit new crimes and do not return to prison contrasts remarkably, and admirably, with the typical situation for the US—within ten years of release, about 85% of former prisoners are convicted of one or more new crimes.

## Portugal Re-Mixes its Failing Approach to Drugs: From 1990 to 2020

In Portugal and the US in the 1990s, about half of prisoners were incarcerated for drug-related crimes. By 2020, that statistic still holds true for the US. But a dramatic re-mix of conditions in Portugal led to far fewer street crimes around drugs and far fewer imprisonments for drug activities.

Back in the 1990s, Portugal was in the grip of heroin addiction. About 1% of the population from all backgrounds—bankers, students, laborers, lawyers, socialites—were hooked on heroin. In connection with this, health problems as well as criminal acts under the legal codes then were abundant.

Government policy in Portugal as well as in the US was counterproductive, consisting of a toxic mix that persisted from the 1970s. Billions of dollars were invested in propaganda demonizing drug use, in police cracking down on drug users, in disproportionate convictions of those from minority and/or impover- ished backgrounds, and in harsh penalties both for drug dealers and drug users. Across the Atlantic, the US was doing the same: spending billions of dollars crack- ing down on drug users.

But in 2001, Portugal introduced a dramatic tricky re-mix. It became the first country in the world to decriminalize the consumption of all drugs. Mixed into the implementation of that decision was not only elimination of huge numbers of arrests for private drug use, but systematic social support for former addicts. Also, attitudes of helping in multiple ways by multiple professional groups were promoted. Anyone caught with less than a ten-day supply of any drug—including marijuana and heroin—is referred to a local commission consisting of a doctor, lawyer, and social worker. Some temporary housing assistance is provided to those in drug rehab treatment. As the positive consequences of all of the preceding conditions cascaded across time, former drug-ridden and crime-ridden neighborhoods were reformed, and many former addicts progressed to no dependence on drugs and to improvements in their careers and social lives. Moreover, further positive cascading consequences have been lower drug overdoses, fewer HIV cases, and fewer overall homicides.

Prison populations and prison costs have declined. Percentages of incarcerated persons for drug-related crimes declined from around 50% in the 1990s to about 20–21% in 2019 and 2020.

In sum, the bold, risky Portugal re-mix on drugs has been a high success in multiple criteria. Many other countries have been consulting with those central to the reforms in Portugal with the hope that, despite cultural and political contrasts, new, successful re-mixes can be achieved more broadly.

## Teaching Children & Adolescents to Resist Drugs

We will briefly introduce here a case example where a "perfect storm" of conditions were mixed together for decades in a failed attempt to head off late adolescent and early-adult drug use. The simplistic ideas which held sway were that drug use is bad and that with enough explanations, games, activities, and attempts at persuasion, individuals will learn to make firm choices against using any drugs. "Just say no!" was the motto, the name of a board game, and centerpiece of countless propaganda pieces.

D.A.R.E. is an acronym that stands for the program most relevant here: Drug Abuse Resistance Education. Developed primarily by law enforcement officials in 1983, the program was provided to children grades one through six and to adolescents through educational curricula. Initiated in Los Angeles in the early 1980s, it expanded across the nation quite rapidly with the passing of the Drug-Free Schools and Communities Act in 1986 by the United States Congress. By the late 1990s, it was reaching about two-thirds of all US school districts. It was heavily funded by the US and state governments, reaching expenditures at one point of over $10 million per year.

We have already seen that without frequent and appropriate monitoring and adjustments to complex plans, the complexity of human dynamics will often veer plans off track and keep going in unproductive directions. This is precisely what happened with D.A.R.E.

One end result was that rigorous evaluation was slow to arrive. But between 1994 and 2002, the research outcomes accumulated, and they showed disaster! The program was not reducing drug use, and in many cases, it was increasing the levels of alcohol, tobacco, and other drug use.

Later, we will explore this "perfect storm" phenomena further. For now, though, we will sum up some of the key conditions involved. The D.A.R.E. delusional tricky mix included the dynamic convergence of at least the conditions below:

- Simplistic belief that law-enforcement officials are ideal educators on drugs
- Simplistic belief that "Just say no!" will work as a guiding light for youth
- Simplistic belief that any child or adolescent who is fully informed of all dangers of drug use will be horrified and avoid them
- Political expediency where hundreds of foundation, state, and federal officials wanted to be part of a popular bandwagon program
- Lavish funding which both supported wide D.A.R.E deployment and helped fuel bandwagon overconfidence
- Hopes of success and overconfidence blinded adult parties to the need for timely and rigorous evaluation—it was so easy and popular to just say "Of course it's gonna work!"
- Failures to appropriately model behaviors by police officers—"I saw our D.A.R.E officer smoking in the school parking lot."
- Limited abilities of many officers to inspire and/or communicate clearly

There are many overlaps between the failures in D.A.R.E and other situations where a hopeful and fairly detailed curricula has been presented to children, adolescents, and young adults on other topics. There is a strong temptation to believe that when well-intentioned and well-informed "experts" develop and present a detailed curriculum it will definitely work without dynamic monitoring and adjustments.

Yet, many children make absolutely minimal progress across five to six years of schooling in math, science, reading, and writing—some of these instances we have reviewed elsewhere in this book. Achieving an effective mix requires more than the instructor simply treating their job as "complete" when they have presented all the planned curricular information. Instead, it is necessary to find enough tricky adjustments and re-mixing to ensure that each student is emotionally engaged, free from anxiety, connecting to the instructor, thinking actively, and actually making progress in comprehending, remembering, and using the information and skills that are targeted.

Here is a quote from vice.com from a man who endured D.A.R.E as an adolescent and became an adult addicted to multiple drugs. His account lays bare how wrongly the processes within D.A.R.E often unfolded:

"We all sat on the floor while people demonstrated people smoking, drinking, and doing drugs. It scared me a little. I was afraid that maybe the first time I tried anything I would die. They told us how addictive crack, cocaine, and heroin are—the statistics, the numbers, the probability of becoming addicted after the first hit or dose. You know what that made me wanna do? Smoke crack, snort cocaine, and do a dose of heroin to see if I could beat the statistics. I did it all."

Chapter 27

# CREATING REMARKABLE DYNAMIC EVENTS

Complexity, complexity, complexity! Many factors need to come together in a timely fashion for significant progress! These themes may seem sometimes overwhelming, to place us in gridlock, as if it would be futile to hope for our own actions and strategies to lead to outcomes we care about.

Yet, we have seen time and time again in the prior chapters how mindful observations and monitoring are able to guide successes despite inherent complexities.

In this chapter, we visit new examples—including some once-in-a-lifetime events—which further illustrate the emergence of remarkable events that transcend our feelings of uncertainty or futility.

## Take A "Pointless" Walk

Many people take walks for many serious reasons. Enhanced cardiovascular fitness or enhanced overall health. To get to a shop or job. To meet a friend. To blow off some steam. To pick up a child from daycare. To collect mushrooms or flowers. To log 6 miles. And so on.

But there are walks or hikes that are entered with no preset goal or endpoint or expectation. These are Zen walks, where the walker relaxes their mind and opens their senses and spirit to whatever may emerge. Paradoxically, such walks often lead to the most intense and satisfying experiences. First, we will examine some of these remarkable walks, both from my own experience and that of others. Then we will revisit dynamic tricky-mixing processes that shed light on why the outcomes are so often exciting, illuminating, and even life-changing.

The remarkable, award-winning poet Mary Oliver has written a number of poems that celebrate exactly these open kinds of walks. Here is an excerpt from her poem "Am I Not Among the Early Risers?":

> "Am I not among the early risers
> and the long-distance walkers?
>
> . . . . . . . . . . . . . . . . . . . . . . . . . . . . . . . . .
>
> Have I not thought, for years, what it would be
> worthy to do, and then gone off, barefoot and with a silver pail,
> to gather blueberries,
> thus coming, as I think, upon a right answer?"
> —New and Selected Poems, Volume Two

## Walking As a Child with a Zen Mentor

Who becomes an adult who shows stewardship and engagement with nature? Research around the world establishes that the most common source of such life-long commitment to nature lies in regular childhood excursions where a parent or grandparent or teacher led the child into natural areas and stoked the child's awareness, openness, discoveries, caring, and excitement. Many of the experiences shared by a child and their mentor are unforgettable across the lifespan. Who can forget their first experience where a "log" floating in a quiet swamp transforms into an alligator who quickly becomes a stunning presence on a bank, soaking in the welcome sun near the viewer?

The long-lasting effects of such experiences often become part of a positive tricky mix sleeper trajectory. In such cases, as a child continues development into young adulthood and beyond, related new experiences of nature cascade into more and more awareness and a coupling to an adult sense of identity. The sleeper or delayed effect will often be a firm sense of adult stewardship toward natural places, creatures, and processes. In the present chapter, some of the experiences of the adults fit within such stewardship. In later chapters, this kind of longitudinal pattern will be further explored.

## My Own Walking into the Deep Woods

Most times, I prefer walking with at least one companion, dog and/or human. But here's a solo walk that turned out to be fabulous.

A little background helps frame this experience. We need to remember that the way we usually chop up our conceptual view of the world misses many opportunities for sensing, appreciating, and having dynamic interactions with the things around us. Objects that seem distinct blur and blend and transform. At what point in their movement and at what distance from the viewer does a flock of one hundred starlings appear as separate birds, and when as one cloud-like single organism? When a pan of liquid cake batter goes into the oven, when and how does it achieve the texture and unity of a cake ready for icing and eating? If we create, admire, and then chomp down a cookie, what do we sense and understand of the cookie that remains within us? When variations in lighting and form and repeated viewing cause us to see alternating "objects" in the same scene, perhaps a vase versus a human form versus a spiritual, other-worldly being, how much do these "illusions" tell us about the nature of what we are doing behind the scenes when the world seems stable, familiar, and easy to label? If the moon does not typically disappear in minutes when in the middle of the sky, how do we integrate our experience of usual moon behavior with that of the moon suddenly being "bitten into" and then occluded by a dark, eclipsing disk?

What we can do to promote the chances of a breakthrough experience is enter the natural world with these conditions actively mixing and converging: keen visual and auditory awareness, leaving behind any concerns or anxieties, an expectation that new events will surprise and delight us, an openness of our spirit to the spirits of wild things, gratitude for the richness and wonders of nature, and entering into the present, the "now," in all the ways we can muster.

*Lynx rufus*, the bobcat, has proven an especially valuable shaman to me on these issues. In Pennsylvania woods I have seen a bobcat on seven different occasions, and each was a sudden and brief exhilarating flash of amazing speed, strength, and mystery. One such occasion was a solo walk in a nearby patch of woods. It was a winter's February morning, and the snow had been falling steadily all through the night in gorgeous, twinkling flakes. There was no wind, and the snow had piled in high, fantastic shapes on every tree limb, rock, and fallen log. It was slow but satisfying work to trod through deep snow across a broad meadow and into the edge of a thick grove of white pines. The quiet was deafening, as if all forest creatures were asleep and the wind had also taken a siesta. My senses were full of whiteness all around, the drifted snow's gentle catch and absorption of each slow bootstep, snowflakes falling close all around my face, some flakes choosing to land on cheek or nose and edging into a soft and very gentle meltdown. The branches, especially those of the dark-green pines, dipped heavily under the weight of their snow sculptures.

In order to move forward through these pines, I adopted the strategy of moving at glacial speed as I edged past each snowy bough without dislodging its snow into my face or down my neck. At this slow pace, in the wind-free interior of the forest, I felt I had never seen or felt any greater peace or beauty in nature. I was accepted by the snow, both in its fallen and now-falling states, by the densely covered trees, and by the few spaces left between boughs. In a way, I was a small, snow-covered tree like all those around me, with the perhaps trivial difference that from time to time I would take a small step forward or lean an arm or shoulder a bit to open a passage beyond the closest appendage of a tree.

I certainly had no inkling of what stood just three steps beyond me as I once again opened a small window and looked beyond the closest heavy arm of a pine. A bobcat! A frozen bobcat! That is, frozen in her tracks, as completely shocked at seeing me as I was at seeing her. In consequence, I could take in and hold forever her whole remarkable shape as she stood in the snow. The black-tipped tufts on the ears and the thin black lines on the face give a sharp, bold contrast with the surrounding whites of this snowy scene. The face is overall a mix of brown and grey, with long whiskers that are pure white and overdramatically long. A dramatic and completely still face. This was unmistakably the face of a bobcat, rather than a lynx or any other cat. The body was sleek, glowing with health, and poised with heavy shoulder and thigh muscles to make some powerful move.

Pause, pause, and then she was off running at the best speed possible through all this snow. Her short tail—dare I say cute, or at least less savage than the rest of this predator—stands out also, with its topping of black, bouncing as she bounds through the snow, creating a floury storm around her. She soon finds a long-fallen tree to ascend, an old bare pine leaning at forty-five degrees against a tall oak. Snow spews, sprays, and falls from this log as she rushes upwards, heading for the high but leaf-bare limbs of the oak. Her speed under these snowy conditions is remarkable, but far less than that of a bobcat on open, dry ground. I am a in position to watch each of her strong strides on her forty-five-degree log ramp, and I see her ever so briefly in the oak.

And then she is gone! I walk all around the oak, probably looking much like a coonhound who is sure that the raccoon it treed remains somewhere up there. But there is no movement, no trace of bobcat or of any tracks of departure. No other tree leans close for an escape route. The beautiful frozen, white crystals continue to fall in complete silence. My vision gets blurry from trying to find a bobcat face or tail or leg somewhere in the irregular, high branches, these branches now a pale mix of grey and brown and white—bobcat colors. This complete disappearance of a bobcat who had stood so close to me before she dashed for the tree is nearly incomprehensible.

Perhaps this bobcat was not real. Perhaps, like the boy in the beginning of the *Velveteen Rabbit*, I did not love her enough or believe in her enough, and so she had not yet become real in the sense that living physical objects continue to exist, to be alive, and be detectable by the usual senses. Perhaps, as it suited her mood and purposes, she could transform from snowflakes into bobcat and then back into snowflakes? Or from tree bark into cat and back again? Perhaps my responses—the degree to which I loved, cared for, wished good-living to the bobcat vision I so vividly saw—might affect its future instantiation?

It is difficult to know for sure. But somehow these events by and in the oak tree must be put together with a couple of subsequent observations. First, a day later, I went back to the spot of the bobcat's disappearance and found unmistakable adult bobcat tracks in the snow not far from the disappearance oak, but not at its base. My inference was that the bobcat had descended to a few feet above the forest floor and then gracefully leaped the final distance and landed like a nimble dancer. Second, a month later, there were many bobcat tracks near a den-like hole in the ground, perhaps two minutes bobcat walking time from the oak. A new snow had fallen, and the tracks were extremely clear along some fallen logs and on certain patches of ground. There were *three* lines of parallel tracks where the tracks were at their clearest. In the center, an adult bobcat walked at an unhurried pace. On each side there were much smaller tracks. Two bobcat kits were out for a stroll with their mother!

For viewing these tracks of the bobcat family, I happened to have with me a great friend, Stephen, who proved witness to what I have reported and who was just as thrilled as I to see that these woods were supporting a new generation of bobcats. It seems clear to the two of us that we had some pretty solid convergence on the reality of local bobcats. Nevertheless, at the same time, we must all maintain awareness of how often the boundaries blur between apparent definite reality and other similar constructions of our senses, hearts, spirits, hopes, and minds.

## Walking From Solid Land onto a Threshold Pier Over the Ocean Edge

I enjoyed lunch with friends in a coastal California café with a deck and an expansive view of the Pacific Ocean on a bright, sunny, windy day. Then I moved on to go walking in the scintillating sun along an adjacent very high and airy old wooden pier. I was walking in pelican style, each step lofted clumsily into air but then slowly, dynamically adjusted on the down-flight to a decent landing. Meanwhile, because of its slow inefficiency, each pace was allowing me a sumptuous look around at gulls and terns and clouds and waves. Beyond this, my absorption in this stroll was richly expanded by the creature to my left matching step-by-step my peculiar ambulation—a true pelican-style walker. She was, in truth, a fine, gray, wildly alive pelican with a fascination with me that answered mine with hers. An unexpected funny yet deep bond held between me and this pelican. I had no desire to be anywhere else on earth doing anything other than this long duet in close concert with my friend.

A skipping, turning, joyful girl bounds up the pier, oblivious to us until she is four feet away and she freezes in her tracks in astonishment.

This young kid goes, "Is that your pelican?"

And so I respond, "No, but I think she likes me. And I certainly like her. She enjoys watching me and walking along with me."

Believe me, that small girl was not the only person astonished by the relaxed, swinging, grooving stride-by-stride waddle-walk that this pelican and I shared for a good long way. We paraded three hundred feet or so down this salty, heavily weathered pier. Onlookers gawked, smiled, and laughed. I was then, and still remain, in awe and at the same time find that this event nourishes some rich reflections.

The first reflection is that I could have improved upon my answer when the girl blurted "Is that your pelican?" Maybe, for example, I could have said, "You bet your boots, my sweet friend, that this is my pelican. But she is also your pelican. This cool pelican belongs to everyone on this pier. She belongs to everyone in the world. She belongs to all the creatures that fly in the air or swim in the sea. So don't you think we better be sure to look out for her, make sure she and her pelican friends have a great life?"

This small episode of pelican-and-person co-ambulation down the high, sea-scented pier also reminds us that when initial expectations are put aside and when we re-mix other conditions on a simple walk, some marvelous breakthrough events emerge. For my part, I initially looked at that pelican taking her generous space on the pier as she dried her wings with a degree of caution. It's very likely from her posture that I evoked similar initial caution in her. We had low and partly negative expectations, in short. ("Do I need to be ready to move quickly around this strange creature?") But with just a short period of quiet, mutual watching, pelican and man dropped their blinders. We allowed a complex and unexpected tricky mix of acceptance; generous, attentive, and fearless behaviors by both; high attentiveness; and enjoyment of the "flowing" perceptions, movement, and mutual regard taking place for hundreds of yards. Across species and size, we felt together the thrum-thrum-pum of the universe, locked hearts, shared a deep connectivity, and began to pulse and stroll to ancient rhythms. We were, for a stretch, part of the long-lasting "animal connection" that the anthropologist Pat Shipman claims, in her book by the same name, goes far back to the earliest stages of Homo Sapiens evolution.

## Two Close Encounters with Mountain Lions

Very early in this book it was revealed that you would learn about how not to be eaten by a mountain lion. Well, here we go! Now we come to stories about meeting a mountain lion at close range. One of these is my story and the other is by a biologist and wilderness adventure leader named Craig Childs. Childs is the author of the book *The Animal Dialogues: Uncommon Encounters in the Wild.*

### My Cougar

The tricky mix of conditions which arose surprisingly one spring day at our Eagle Spirits Farm is here discussed. As in so many other tricky mix events, this event is unlikely to be repeated anytime in my future.

The cougar (also called a mountain lion) and I met face-to-face at a distance of only about thirty feet. Here are the conditions of the mix that caused the mountain lion to be where he was and led me to be approaching his position without his sensing me until I was quite close.

- The hay in the meadows is two feet tall because the spring has been wet and cool—perfect hay-growing weather
- The hay is this tall also because weather had prevented cutting the hay up until this point
- I was walking without the company of any other person
- My companions were three large Labrador retrievers, buried from sight by the tall hay
- The direction of the wind carried his scent toward me but did not carry my scent or the scent of my companions toward the mountain lion
- The tall grass hay had recently harbored multiple deer, and even as the dogs and I walked through the hay it was likely that somewhere in the meadows one or more young deer were lying quietly
- My cougar was moving down from the mountain into the meadows because he planned to hunt the deer

So the convergence of just these conditions created the mix whereby my movement from the house and barn in the valley upward toward the mountain and the mountain lion's path of movement down toward the meadows directly collided.

Then the mountain lion stopped dead in his tracks and stared directly at me. On my part, I was astonished and also stopped dead, kept quiet, and keenly riveted my attention on this amazing creature. His legs were shorter than you might expect for a critter whose mass and weight were probably four times that of one of my dogs. In contrast, his shoulders were massively muscled and taught. His remarkably long tail stretched behind his body—a tail that helps mountain lions take great bounds and leap at high speed onto a branch twenty feet off the ground. His eyes glowed at me, and his entire expression said, "How dare you bring your body into the path of my ongoing venison hunt!"

Something had to give on his part or my part. Fortunately, he decided not to eat me nor to continue his plan to search the meadow thoroughly for deer. Mixing into the ongoing tricky mix of events were my decisions to hold my ground, appear tense and ready for conflict but not fearful, and to in no way wave, yell, or otherwise threaten the mountain lion.

Breakthrough! The total mix of conditions proved positive and tricky enough for a favorable resolution. My mountain lion turned in a deliberate fashion and began to walk slowly through the grass, but now in the direction of the mountain, moving with disdain and with not a single backward glance or pause. He just gradually melted from view into the tall grasses.

*Looking into the Eyes of a Wild and Free Mountain Lion.*

## Craig Child's Mountain Lion

Many of the conditions we have just discussed in the case of my mountain lion also applied in a very dramatic situation which arose between Craig Child and a mountain lion in an arid section of the American West. But there are also a few very important new twists or tricks which arose—key components of the tricky mix there.

- Like me, Child was moving along out in nature with no human companion with him
- Craig's background included incredibly extensive experiences from childhood on in natural places
- His background included professionally leading wilderness trips, which acquainted many clients with the wonders, surprises, and complexities of wild creatures
- He had already safely escaped quite a few harrowing incidents in wild places
- These background elements together led him to a dangerous level of self-confidence
- He therefore set out on a particular trek across the desert and on a difficult climb onto a high mesa
- This climb resulted in quite a few minor scratches and cuts which were no real threat to him but nevertheless meant that blood would arouse the keen sense of predators such as a mountain lion
- Craig's decision to travel light on this long trek meant that he would rely on some water available in a small pool in a ravine
- He dismissed the brief sighting of a mountain lion in the ravine, a lion who fairly quickly moved away as an indication that the lion was eager to get as far away from him as possible

THEN, in this isolated spot, this solo human traveler found that as he drew near to the water for a cool drink the mountain lion reappears quite close—only about thirty to forty feet away. What happens next is a complex remixing of conditions which could have gone either way quite easily—towards Craig being eaten or toward his safety when the lion chooses to leave. So let's see how the further re-mixing unfolds

- Child is deeply knowledgeable about the behaviors and tendencies of mountain lions and persistently manages to keep facing the lion directly rather than ever, even for a brief moment, turning his back to the lion
- The lion has a history of almost always taking down his prey by attacking it from behind and quickly crunching their skull with his sharp teeth and powerful jaws. Thus, he repeatedly moves to the to try to get behind Childs, only to see Childs smoothly move to constantly face him
- The mountain lion re-mixes an attack plan and chooses to slowly step closer to Child's position until he stands a mere ten to twelve feet away
- This lone man also re-mixes his plan, by bringing out the only meager weapon at his disposal, his one knife, while maintaining his steady gaze on the lion and a firm and confident stance

Okay. So it came down to re-mixed tricky mix lion versus re-mixed tricky mix man. What happened in the end was strikingly similar to what happened with me and that mountain lion so close to me in our grassy meadow.

The mountain lion decided to retreat and slowly walked away down to the mouth of the ravine and out into the desert, very gradually indeed, disappearing from view without a single glance back at Craig Childs.

## Closing Thoughts on Connecting Persons Mutually to Wild Animals, and Particularly Concerning Unexpected New Connections

In the words of the Buddhist scholar and teacher Thich Nhat Hanh in his book *Mindfulness*:

"When we look at the blue sky, the white clouds, and the sea, we are prone to seeing them as three separate phenomena. But if we look more carefully, we can see that the three are of the same nature and cannot exist independently of each other . . . The mind and the world contain each other so completely and perfectly that we call this 'perfect unity of mind and object.'"

# Chapter 28

# DEAL-BREAKERS, NAYSAYERS, WRECKERS, AND OTHER OPPONENTS

Oftentimes, after a considerable period of thoughtful effort and adjustment, a great many seemingly favorable conditions are brought together to try to support progress.

Yet, for an individual as well as for a corporate, educational, or other entity, substantial progress remains tantalizingly out of reach.

In this chapter, we first examine circumstances where a single dissonant factor effectively serves to undermine the needed dynamic convergence of conditions for success. Such a factor may be called a "deal-breaker."

## Inclusive Preschools for Immigrant Children

In Sweden and the USA, among other countries, a shared goal of many parents and educators has been to prepare immigrant children for schooling in the native language of the new country. If a child at three or four years of age who speaks a different language at home were to learn at preschool the country's dominant language, then that achievement would be a stellar step up toward success in school at six-or-so years of age.

I have observed this personally and also reviewed research reports relevant to these possibilities. What I have seen is repeated deal-breakers that served to undermine success despite what had seemed an overall promising preschool approach.

First, we look at the key factors in multiple countries that were seemingly building toward enough convergence of conditions to bring immigrant children's language levels up to the local norms in their new country.

1. Immigrant children are mixed together with native language speakers who are adult teachers and aides
2. Immigrant children are mixed together with native language speakers who are local children, and these local children make up 50–67% of the preschool classes
3. The preschool's planned curriculum provides many hours per day of relatively unstructured play activities where extensive conversations would be possible in the local language
4. The parents of the immigrant children are accepting of the goal of their children acquiring the local language, yet proud of their own language with their children at home and proud of their children's high age-appropriate competency in that language

In combination, the above conditions would appear to provide a preschool setting in which the immigrant children would likely encounter many highly fluent examples of how local children and adults employ the language.

Yes, the overall preschool plan indeed looked promising. After all, if the immigrant children were going to be in a preschool anyway, why not one where there is an opportunity for plenty of exposure to the local language?

But here's the crucial deal-breaker. For the most part, the potential for lots of daily conversation involving the immigrant children engaged with local speakers was *not realized*. Despite expectations to the contrary, most of the Swedish conversations in Sweden and the English conversations in the USA excluded the immigrant children. Socially, the immigrant children primarily talked with each other rather than to local adults and children.

Children who are new to a country with no prior knowledge of the local language at age four *are capable* of rapidly making so much progress that they catch up to local children in just one year—by age five. Other research establishes that conclusion. But the children catch up only if a full set—a full tricky mix—of supportive learning conditions converge. Designers of the above preschool programs, which failed miserably in going beyond the mere physical presence of a mix of immigrant and local children, took steps that would ensure extensive and positive conversational interactions between the native language speakers and the immigrant children as learners of that local language. To carry out that latter reality would not require additional money. It would instead require respect for the *complexity* of conversational language learning in meaningful and emotionally/socially positive daily interactions. In short, it would require re-mixing social interactions between immigrant children and native speakers.

## We Were So Close to Fabulous Progress and Then . . .

In project teams that have worked together for extensive periods across multiple projects and years, even when the dynamics so far have been excellent, sudden deal-breakers may appear.

There is a paradox inherent in long-term work or personal relationships. The longer and more open the relationship, the more each party knows about the other. By similar dynamic processes, such knowledge, such intimacy, can support sensitive and empathic mutual adjustments. And yet, if some new event triggers anger or suspicion or mistrust, then that same knowledge about the other may lead one to an attack that strikes at the other's known vulnerabilities, self-concept, and personal image.

Under these circumstances, the deal-breaker that leads to group dysfunction, or even dissolution, despite a long, positive history together can be a one-time insult or betrayal.

## Pandemics: Public Health Dynamic Tricky Mixes Which Cascade Positively Toward Favorable Progress or Instead Encounter Wreckers and Extreme Negative Cascades

There is no possibility of escaping the consequences of dealing with Dynamic Systems through contrasting action patterns.

The same basic, underlying dynamic convergent processes will be at work when projects play out toward high successes or toward stunning failures. What makes the difference is the particular set of tricky mix decisions and actions which are established, including what monitoring strategies are implemented.

In the case of the COVID-19 virus pandemic of 2019–2021, a powerful "starter kit" of components to include in a mix to try to control pandemic effects had been well established before COVID-19 appeared. Here is that starter tricky mix in capsule form:

- Frequent infection-detection tests are performed
- Labs work on improving testing speed and quality
- All individuals with positive tests results and/or common symptoms are quarantined as well as recent persons in contact
- Contact tracing is systematic
- Health workers are given state-of-the-art protective equipment and procedures
- Masks are worn during all social contacts outside the home
- Size of social gatherings is severely limited
- Effective monitoring strategies are in place
- Compliance with these public plans is considered an ethical imperative by all involved
- Actual compliance is nearly universal

## Case Study: Taiwan

In Taiwan, government officials and a responsive public rapidly put in place the above starter tricky mix and then tweaked its implementation depending upon each week's monitoring of how well each component was playing out.

Despite its close proximity to the China mainland and risks of the virus spreading from China, Taiwan has achieved exemplary success in the management of COVID-19. As of March 2021, in a country of about twenty-two million, only 1,007 total cases and just ten virus-related deaths have occurred!

Similar levels of success in limiting COVID-19 impacts will be addressed in Chapter 36 regarding Dynamic Puzzle #12.

## Case Study: Wreckers in the USA Create a Cascading Catastrophe

By April 2021, in the USA, over 550,000 deaths had occurred, and there is no end in sight. A staggering thirty-one million Americans have been infected.

Dynamic tricky mix processes have been playing out for sure, but primarily in negatively cascading directions. Laying out any of the known starter tricky mix conditions was dangerously slow and inconsistent. Testing was lax rather than aggressive, and contact tracing was irresponsibly slow to develop and still not widely in place. Worse still, even in communities which are documented "hotspots" of infection, mask-wearing, limits on social gatherings, and provision of adequate protections for health workers are often lacking.

Obvious keys to these massive failures are the roles that multiple "wreckers" have played. In combination, these persons in positions of high responsibility, by their actions and their inactions, comprise a new toxic leadership mix.

By their executive decisions and their public statements, these leaders have misled our society and undermined the needed convergence of all the components which Taiwan and multiple other countries have shown can be implemented in coordinated, timely fashion. Perhaps most damning of all, these wreckers have persuaded a large number of Americans to treat defiance of protective actions as an "ethical responsibility." Such misguided defiance then dynamically serves to disrupt needed monitoring and needed ratcheting-up of all the known important conditions for combatting a pandemic.

Others will make thorough efforts to recount who did what, when, and for what motives. All of this may help nudge future decisions on public health matters toward responsibility, creativity, and effectiveness.

But as of March/April 2021, at a minimum, it is clear that central, influential, persistent wreckers include multiple Republican governors and senators. Most influential of all in setting up and continuing toxic negative counterproductive mixes was President Donald J. Trump while he was in office.

Chapter 29

# HUMOR AS A CATALYST FOR NEW EFFECTIVE MIXES FOR SMALL BUT ANNOYING PROBLEMS AS WELL AS LARGER ISSUES

Each of us faces times when we have to figure out what to do about a broken-down refrigerator, car, cell phone, and the like. Usually, we lack the expertise to make good, quick decisions about what to do to make things right again. Simple strategy—turn to a familiar business or friend and trust them. In such circumstances, a more effective but more complex strategy is to also seek out and integrate a range of independent sources of information and then decide.

The simple trusting approach relies on our fairly "local" knowledge and networks. It is extremely easy to be led down the wrong path if we have too much confidence in our local situations and sources. Many variations on how economically important judgments of this sort can go wrong have been richly detailed and modeled by Nobel Prize winners Tversky and Kahnemann for individuals as well as for corporations and countries. They strongly remind us to keep looking until we actually understand more about complexity.

Part of such exploration of possibilities is to make contact with others who are able to jolt our thinking and openness to new ideas by invoking humor.

## Humor in the Context of Cars

A very different pair of collaborators have achieved fame in the domains of vehicle performance and vehicle repair. To see how some folks have dealt with their puzzling car or truck troubles, let's now turn to a source of expert information delivered with high flourishes of humor (which in turn facilitates good judgment). That source of information is the Tappett Brothers, stars of National Public Radio's *Car Talk*. To get to the essence of the dialog between troubled vehicle owners and Tom and Ray Magliozzi, consider these excerpts.

Question by Bob:

> I would appreciate your opinion on the problem I had with my Ford F-150. An advisor called to say my battery would not take the charge and that it appeared as though something had wiped out my battery, voltage regulator, alternator, and starter. It cost me $686.00. Does this sound plausible?
>
> Ray: Gee, Bob. It's highly unlikely that your battery, voltage regulator, and starter would all go bad at the same time by themselves. I mean, the chances of that are about the same as the chances someone would greet my brother as "Hey, handsome."
>
> Tom: In the interest of vile innuendo, . . . here is a possible scenario. Your truck is towed . . . and the next morning Curly the mechanic tries to start the truck and finds that your battery is weak. He tells Shemp, the nineteen-year-old garage gopher, to hook up the jumper cables. Shemp has three jobs at the garage. He sweeps the floors, he gets coffee for the other mechanics, and he jump-starts cars. The sweeping floors is the only job he seems to have gotten the hang of.
>
> Ray: Yep, so Shemp hooks up the jumper cables backwards and fries the whole works. Then he plays Mickey the Dunce and you end up paying $686.00.

New question from Michelle:

Since this is my first time owning leather seats, I'd like to know if there are any special techniques to care for my leather upholstery. Any suggestions?

Ray: Actually, there is not much you have to do with leather. That's its great advantage. I mean, you don't see cows wiping each other down with Armor All Surface Protector out in the pasture, do you?

## Larger Issues, Such as National Security, Scientific Breakthroughs, and Long-Term Mental Health

Some issues may at times appear to us as just too vastly important to allow any humor to enter in.

Nobel Prize winner and physicist Richard Feynman finds that position absurd. In his view, humor and pranks loosen up our thinking, help to open our eyes, and build positive social bonds between those involved on a project. Personally, it is evident he also just has a lot of fun experiences, some of which are covered in his two popular books, *"Surely You're Joking, Mr. Feynman!": Adventures of a Curious Character* and *"What Do You Care What Other People Think?": Further Adventures of a Curious Character.*

His work on US national defense projects required some ways that research and development teams could achieve very high security for the project documents. Feynman's pranks opened responsible security folks to the limits of what they had in place. For example, for fun, he learned how to pick locks on doors and file cabinets and even how to "tumble open" complex combination locks on large safes full of secrets. By demonstrating brazenly that he could quickly break in and see the files of other project members, he forced them to re-mix their security plans to better defend against enemies of the project and the USA.

Similarly, a broad range of scientists including Feynman and Niels Bohr have claimed that the key breakthroughs they enjoyed during collaborative research were richly supported by team processes that incorporated frequent humorous events.

Recall, also, that in Chapter 21 we became acquainted with the research of George Vaillant concerning long-term patterns of development for individuals. Humor was one essential component of longitudinal outcomes. Those men who employed humor effectively to help cope with whatever problems and issues arose between age twenty and age seventy-five were more likely to cope better than those who did not. For men with the most successful careers, highest mental health, and best physical health over the long run, humor was a key part of a larger positive tricky mix.

## Why Cats Paint

Hmm. You may be thinking, "Gee, I never even knew that cats painted." Well, you will see below some rare—humor-inspired—examples of cat artists and their colorful creations.

*Sample of humorous cats as artists in book Why Cats Paint: A Theory of Feline Aesthetics by Heather Busch and Burton Silver.*

A broader lesson or two may be drawn (haha) from these paintings by cats. First, if we have *never* run across an example of an event/achievement before, we are wise to be highly skeptical. But, at the same time, we need to maintain some openness and some curiosity about whether, under newly understood dynamic mixes of conditions, the claimed achievement would ever occur. As just one relevant example, we now have multiple frameworks for making some sense of fossils from creatures dating back many millions of years. Yet, three hundred years ago, what we now interpret as fossils were pieces of rock that were totally baffling—and any speculations about an ancient past were treated as impossible narratives.

In the case of cats, my hunch is that, regardless of new conditions, the limitations of their brains will continue to preclude intentional, complex, aesthetic art productions.

A related question for humans is: Why do most children *not* paint at any complex level? In the past, some observers have been too quick to jump to the simplistic conclusion that most children lack the "art gift" or "art brain" required. In our next chapter, a more adequate account is couched in Dynamic Tricky Mix theory.

## Summing Up

In summary, an absence of an active and constructive sense of humor is no laughing matter. That's because humor feeds convergently and powerfully into dynamic mixes for all manners of projects in our private, social, and political lives, and for all fields of science and the arts.

Chapter 30

# DA VINCI'S CHILDREN, DA VINCI'S ANCESTORS

Here we visit recent breakthroughs concerning the potentials of our prehistoric ancestors in art along with tightly interconnected breakthroughs concerning the potentials of modern children—"Da Vinci's children"—in art production and comprehension. Modern children are not, of course, literally da Vinci's children, but *if they are given supportive dynamic mixes akin to those provided to da Vinci,* what they are capable of achieving in art indeed follows his artistic lineage. Moreover, and in keeping with dynamic convergent processes, we will see that a fundamental reworking of accounts of early hominins' use of tools, weapons, and symbols is required also.

The new accounts offered in this chapter bypass multiple limiting biases. The accounts are, as well, integrative of exciting new discoveries, detailed theoretical processes, and a broad range of disciplines. Long-standing puzzles are solved. Be prepared to see the world anew!

## Da Vinci's Ancestors and the Earliest, Earliest Symbols

Preceding any other tangible evidence of symbolic behavior, early hominins collected, transported, and socially shared shells, stones, and fossils. These were among the first symbols of any kind in prehistory, and they required no tools or craftmanship. Here we argue that employment of these found symbols transformed the complexity and precision of communication and thus carried significant value. This chapter further specifies how over time such borrowing of symbols from nature led to multiple breakthrough cascades of strong co-evolutionary advances in brain and culture.

## Limiting Assumptions in Archaeology and Anthropology

It is now clear that a number of arbitrary assumptions in the field of archaeology have warped and distorted many lines of new discoveries and the interpretation of those discoveries:

1. The stones/ artifacts will speak for themselves. That is, archaeologists and others should drastically limit any speculation about the meaning of artifacts, how they were shared, and how they fit into the primate or hominin culture of the time

2. The only artifacts of interest are ones in which it is reasonably clear that intentional marks, cuts, or other imprints upon stone, bone, or other material were achieved.

3. Art within a clan or culture consists only of intentionally made, clear-cut working of material by the presumed prehistoric artist

4. An art structure cannot be interpreted as symbolic unless it is reproduced in different exemplars and unless the symbolic representation is realistic/figurative and includes many features which are obvious to our modern eyes

5. Art could not have been part of hominin culture until after tool use had reached significant levels and had already driven the brain toward higher complexity and size

6. Art at all points in prehistory from about two million years ago to the start of the Upper Paleolithic at around 35,000 years ago was in no way a central contributor to the progress toward other complex behaviors and thinking patterns

The majority of theorists, until recently, have worked within most or all of these assumptions. This has led to a lack of interest or a denial of significance for seeming artifacts with no functional value (non-tools) with dates greater than 40,000 years ago. There has been special, deep skepticism for artifacts from the period of one million years ago to 100,000 years ago. The field of archaeology has only recently provided even a few authors who see the dangers of these assumptions. But in 1997, Bahn and Vertut provided a refreshing exception with these statements:

"The earliest abundance of any form of archaeological evidence should never be interpreted automatically as the earliest occurrence of a phenomena. The further back in time we look, the more truncated, distorted, and imperceptible will the traces of 'art' appear."

In contrast to those comments, Lorblanchet and Bahn sum up the theory that has most often been dominant in the literature and museum displays throughout the world concerning prehistoric art:

"[One theory] which could be called 'revolutionary' because it stresses the break that seems to mark, in the view of its proponents, the arrival of the Upper Paleolithic, seen as a revolution, tends to minimize everything that preceded that period."

The problems for this theoretical position and for the above-mentioned biasing assumptions is that far too much hard physical evidence has, by the present day, emerged to the contrary. The assumptions are all, at best, arbitrary, and most of them have been proven wrong and challenged by new theorists.

So, what we have here is another familiar situation to those analyzed in this book for other domains and situations. Theorists have been *stuck* in their understanding of art in prehistory because their guiding assumptions have blocked them from fully addressing the demonstrated art potentials and achievements of multiple hominin species and very early Homo sapiens.

We now can reveal and interpret previously unexamined, concrete origins of significant symbolic behavior because Da Vinci's earliest ancestors left evidence in plain sight—these ancestors were finding art and making art at 300,000 years ago to two million years ago.

Next, we spell out the essence of recent theoretical breakthroughs that give more adequate accounts of how breakthroughs in symbolic representation arose, including collection and creation of the first art objects that occurred in prehistory.

## Beyond Catch-22 Traps and Circular Accounts

How did an early species that lacked symbolic behavior of any kind begin to use symbols? Circular, inadequate, simplistic accounts posit a lucky, sudden genetic mutation or two that now made the brain a symbolic brain—with no tangible evidence whatsoever that this mutation occurred.

The heart of an adequate theoretical account is that some clans within a certain hominin species, without changing their brains at first, began to use those existing brains in powerful new ways. Just as is the case with you, the reader, some of these very early prehistoric ancestors have open the option to re-mix certain sets of perceptions, plans, and actions. Rather than just perceiving rapidly all manner of complex shapes when hunting for food or places to shelter, they made a "second level" leap to exploiting similarities between newly collected shapes and shapes already well represented in their visual memories.

Rather than simply seeing and gathering a fish, for example, they saw and gathered rocks or shells that resembled fish. Thus, these collected objects could be carried around and shared as symbols of fish. Back in a cave, for example, pointing to the fish rock could initiate sharing a plan about fishing. It is crucial to recognize that a symbolic, socially shared fish rock is a powerful catalyst for expanding communication, and also a piece of representational art. That nature made the art form rather than a hominin creating it does not matter to its significance.

The significance of the fish rock art lies in social process, just as a 21st-century artist's acrylic painting of a fish, turtle, or snake has significance only through socially shared interpretation. For the first clans that began finding and sharing artistic symbols, it is important to note how relatively easy this step would have been for anyone in the clan. That is, once a symbol was collected, the communicative sharing could begin. No breakthroughs in terms of inventing new tools and techniques and training clan members in using these were required.

## Very Old Collected-and-Transported Symbolic Objects

Below you will view three different ancient, collected stone objects (no shaping by tools, etc.) from hominin-occupied sites at varying locations from Morocco to South Africa. These have immediate representational, symbolic value to our modern eyes as well as to those of our prehistoric ancestors at dates between

300,000 to two-and-a-half-million years ago. These symbolic artifacts reveal an essential foundation for art and symbolism that far predated the more famous artistic creations of statuettes, cave paintings, and desert rock paintings discovered from the period of 25,000 to 40,000 years ago.

Portable Symbolic Woman, the 400,000-Year Old
Tan-Tan Venus Figurine from Morocco
(Bednarik image).

Symbolic Penis—Cuttlefish Fossil Transported
to Dwelling Site in Morocco 300,000 Years Ago
(Bednarik image).

*This "Makapansgat Stone" is Prehistory's Oldest Symbol and Oldest Art (Bednarik image).*

Thus, da Vinci's ancestors began the art journey by finding and sharing found art/symbolic specimens. And when they did, the consequences for brain complexity and social complexity were enormous. Moreover, many of these consequences came relatively quickly—in just ten-to-one-hundred generations rather than millions of years.

## New Research Likely to Unearth Far Earlier Prehistoric Symbols/Art Collected by Hominins

It is essential to see that the biases reviewed earlier have placed severe limits on research activities concerning any collected objects of symbolic interest and value from early hominins. Instead, tools and tool-shaped artifacts are what archaeologists have collected and placed proudly in museums. When more researchers look diligently for the unshaped collected symbols, far more examples will emerge from the period between two-and-a-half-million and 100,000 years ago.

The hominin species existing and evolving during that period possessed essentially the same visual systems and motor systems as we modern humans have. Accordingly, once they made choices to look for and transport natural symbols from nature, no physical limits would hold them back.

View now a few natural objects which I have collected. These illustrate how very abundant objects of potential symbolic value and easily portable sizes are in our modern environments as well as in environments all across the last two-and-a-half-million years.

Keep in mind that to work successfully as a symbol, a rock, nut, or other object need not have a full range of features. For example, if two early hominins or two modern humans *communicate agreement on a symbol's meaning*, that is all that is required. Thus, one of the objects pictured below, a fossil "fish rock," can function well as a fish symbol despite having both similarities and dissimilarities to any actual fish. Similarly, an object with a pattern similar only to the head of an animal or person may function efficiently as a symbol for the whole.

*Small, Portable, Fossilized Fish as Symbol of Live Fish as Food Source.*

*Modern Walnut Shell Symbolizes Human.*

*Small, Portable Walnut Shell as Symbol for Owl.*

## Survival Value Drove Genetic Transmission and Cascades of Brain and Behavior Changes

Early on in hominin evolution, the invented use of found objects as socially shared symbols gave the only clans using them impressive gains in survival value. Clans using them as single symbols, and even more so in new combinations of symbols with expanded, precise meanings, would be better in multiple domains. Such domains included at least: avoiding predators, finding shelter, child protection and care, longer lives, more and better-nourished children, and more successful relations with other clans. Within clans, the best of the found-symbol users would enjoy the highest survival and fecundity rates and thus would contribute their genes at the highest rates to later generations.

We argue that fundamental cognitive-social processes in learning to abstract and share symbols in early hominins are very similar to those in young humans at about ten-to-twenty-four-months when they are acquiring their first symbols and symbol combinations (Nelson, 2016; Nelson et al., 2004; Nelson & Arkenberg, 2008; Tomasello, 2008). As brain size advances from 600–1,100 grams, along with accompanying brain complexity, the speed and power of such processes accelerates both in early hominins and in modern developing human infants and toddlers. Moreover, we can see five likely phases once tangible symbols enter early hominin clans.

- Phase one: found symbols expand a small existing repertoire of points, gestures, and pantomime.
- Phase two: some of the found symbols enter combinatorial "phrases" such as "many big fish."
- Phase three: explosive growth of multimodal combinations/phrases driven by increasingly active cultural collection of newfound symbols and increasing awareness by clan members of the value of symbols. Varied multiple-symbol phrases consist of found symbols plus pantomime, found symbols plus pantomime plus gesture, and found symbols plus points plus gestures.
- Phase four: for the first time, after advanced brain size and capacity are in place because of selection and evolutionary advantage driven by the first

three phases, reproducible speech symbols and invented sign symbols enter into clan culture. Speech and sign, although primitive, now contribute symbols to multi-domain combinations that still include found symbols from nature.

- Phase five: based upon more advanced brains and an established foundation of symbol use, sometime after 200,000 years ago new invented symbols are constructed to decorate tools, weapons, clothing, tents/dwellings, pottery, and jewelry. Art creations in all media become pervasive in the most advanced clans. Eventually, at around 10,000 years ago, systematic writing systems begin.

## Sophisticated Art from 25,000 to 35,000 Years Ago in France, Spain, and Austria

Only in the last one-hundred-fifty years have discoveries of depictions of horses, bison, lions, rhinos, and other creatures on the ceilings and walls of caves been recognized as dating as far back in the Paleolithic period as 35,000 years before the present. The artists who made these surviving paintings, including complex scenes and a sense of animated movement, clearly possessed a modern-like tool kit of graphic skills. Without a doubt, for some time before 35,000 years ago, some clans must have been valuing, incorporating, and culturally transmitting a powerful set of drawing, engraving, and painting skills along with a diverse set of artistic products.

Similarly, by 35,000 years ago, some clans in Austria, and more broadly across Eastern Europe, were creating beautiful, small statuettes or "Venus figurine" carved from stone. Many of these were valued highly and were transported to accompany nomadic movements.

Examples of both the small Venus figurines and the cave paintings are shown below. Note, once again, that foundations for these art creations are evident in the prehistoric records going back 400,000 to two-and-a-half-million years ago—tangible, physical, collected, and transported natural objects were influencing cultural symbolic communication and physical evolution.

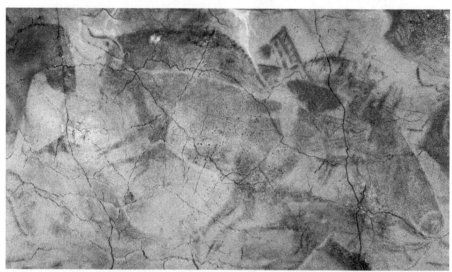

*Altamira, Spain, Cave Painting of Bison.*

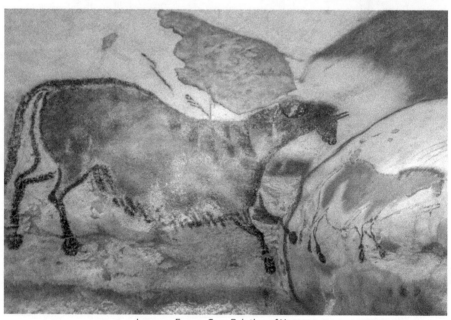

*Lascaux, France, Cave Painting of Horses.*

*Small, Portable Venus of Willendorf Figurine from Austria.*

*30,000-Year-Old Venus Figurine from Stratzig, Austria—Note that Two Views Show This Artist Portrayed Many Features of a Woman's Body.*

## Da Vinci's Children

There are striking parallels between discussions of modern children's art potentials and discussions of prehistoric art by archaeologists. For modern children, the predominant position has been that potential for complex art-making is lacking in most children and that a biological/genetic "gift" underlies the very, very small percentage of children between three and twelve years of age who acquire significant art skills. Similarly, for prehistory throughout the last two hundred years, a core assumption in scientific writing has been that for most of prehistory, hominin brains were biologically unprepared for art-making.

Within our Dynamic Systems approach, we see instead an alternative set of parallels. Although early hominin brains were adequately prepared for art, the leap into finding and using art occurred only when the right shift in a clan's active visual comparisons and the right social dynamics converged; before that, all hominin clans were *stuck* with no art and no visual symbol system of any kind. Likewise, for modern two-to-five-year-olds, their brains are adequately prepared for art-making, but we find that potential realized in complex art production for less than 1% of preschool children. Further, that usually untapped potential yields significant arts skills for children precisely when they encounter a tricky dynamic mix of highly-engaging encounters with skilled artists while the artists and children are both actually making art.

## Solving Another Dynamic Puzzle from Chapter 5

In Chapter 5, a dynamic puzzle was presented related to the art potential of young children. The puzzle to solve was why most four-year-olds are so primitive in their drawing and artistic expressions—the levels of their best drawing efforts lie at three, four, or five on the graphic below of stages three to eight in drawing development. That puzzle is deepened by the fact that most four-year-olds have already demonstrated fantastic learning abilities when it comes to speaking one or more languages—their brains obviously have proven capable of dealing with highly complex, meaningful communication.

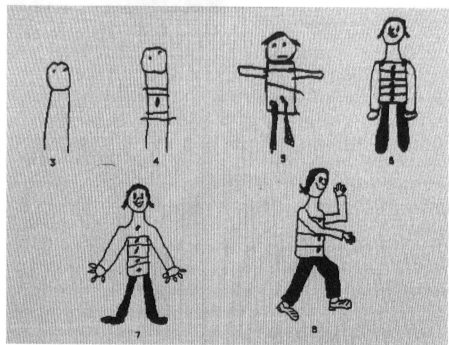

*Six Stages of Advancing Complexity in Children's Drawings of People.*

A powerful explanation rooted in scientific experiments has surprised many theorists, parents, and teachers. In our lab, we have done causal intervention studies in which a mere ten to twenty hours of newly mixed learning opportunities are provided to a four-year-old child who can only draw the simple egg-like person on the left in the graphic below. The child then jumps up by the end of the intervention period to drawing at the complex level pictured on the right.

The first rigorous experiment to see what young children would learn from artists was set up with preschool children. Elizabeth Pemberton and I found artists eager to work with young children and trained them in a nondirective but challenging procedure as described below. We had pretested the children and found before the training period, as expected, a quite limited set of art skills—the children, for example, drew people as simple combinations of circles, dots, and straight lines, without clear portrayal of emotion, movement, or interaction with objects. Accordingly, when they first encountered our procedure, there were plenty of so-far-unlearned art skills that the children could potentially learn from the artists.

*This Child Drew Human Figures with Limited Complexity at Pretest (Left) but Advanced to Make Figures like the One on the Right after Only Fifteen Hours of Interacting with a Skilled and Engaging Adult Artist.*

That first study has now been replicated multiple times in the USA and Sweden with absolutely consistent outcomes—children who at pretest are very primitive in their art advance within fifteen to forty hours of artist/child interaction to much more complex and highly skilled drawings and paintings.

In a way, what the artists did with the children was very similar to what good, well-trained conversationalists do in language with young children. The artists found topics that were of interest to the children, invited the children to draw, and then "conversed" back and forth with the children in art (and in natural language as well). If the child drew a lion, so did the artist, but at their more-advanced skill level. Back and forth the exchanges would go, sometimes building a coherent story and sometimes not, but always leading to fun exchanges in which the child again and again received chances to compare what they built into their drawings with the contrasting level of drawing provided by the artists. The child could see in real-time how the artist began, proceeded, edited, and completed drawings which included techniques—at that moment—beyond the child's own level. At the same time, the child's drawings were accepted as a shared topic, acknowledged and encouraged, and shared in a positive social-emotional series of exchanges. Even though the child was never asked or told to do any new technique that the artist presented, we were gambling that it would be easy for the child to find the artist's many productions interesting and, in some instances, worth trying to emulate.

Let's now sum up the causal experimental studies as well as the creations of the most exceptional child artists within the dynamic tricky mix framework. What we have observed for a very small minority of 20th and 21st-century children is undeniable production of sustained series of artworks at high, interesting levels of complexity, style, and aesthetic appeal. "All" that seems to be required is to completely break the usual mold of *impoverished* artistic opportunities. Complex art skills emerge when, for sustained months or years, fortunate children and early adolescents have these components mixed into their lives:

- Frequent opportunities to interact with multiple skilled artists in socially and emotionally engaging circumstances
- Rich tools and media to work with
- High challenges provided by frequent observation of sophisticated techniques in use by the already-skilled artists at work in real-time

- Recognition and encouragement for advances in their painting, drawing, sculpture and other creations

When all of these components are converging dynamically across several months, tricky mixes are formed which sustain children's progress at *flowing* levels. What the children achieve in art delights and encourages the children as well as their parents and teachers.

Beyond that, there is in extremely rare instances a level of dynamic convergences and positive snowballs which fuel a child's progress at *soaring* levels for multiple years. Let's take a look at the Chinese child Wang Yani and her art at three to five years of age. The examples below show that Yani as a preschool child has already achieved professional-level art creations!

*Yani, Age Three, "Picking Fruit."*

Yani, Age Four, "Performing Acrobatics."

Yani's first picture, with the monkey touching fruit, was painted at age three . She painted the second one, two interacting and actively performing acrobatic monkeys, at age four.

Yani *Soared* toward higher and higher complex art-making skills across the ages of two to six because the mix she enjoyed was worlds apart from the art experienced by most young children. For hours almost every day, she entered her artist-father's large backyard studio and had access to fabulous art materials and observed and interacted with her father and multiple advanced adult art students. They encouraged her and allowed her to see in real-time all the steps they employed in making their own paintings. Contrast all those elements with what is typical for a preschool child anywhere in the world—not even one skilled artist is ever available to the child for interaction with and observation of skilled art-making in action. Moreover, the *intensity* of the components of the tricky mix for Yani were all high. The incredibly rich positive dynamic mix she was immersed in led her to professional, expert levels. Even at age four her paintings were on exhibit in China! Further, across her remaining childhood years she accumulated an impressive, accomplished portfolio of paintings which went on exhibit in museums around the world—Paris, San Francisco, Tokyo, and more.

Implications from the experiments we reviewed and from Yani's case are very similar. Whenever we enrich children's mixes of supportive experiences in interacting with artists and making art alongside them, the children rapidly enlarge

their art tool kit and deepen their interests in making art. If the enriched conditions are maximal and if they continue for multiple years, then any preschool
child likely will show remarkably high, near-professional or professional levels of
skill and accomplishment. This latter outcome is what we see with Yani—the best
documented case so far of a young child's exceptional progress and the remarkable opportunities she had for learning, encouragement, and recognition. Beyond
Yani's case, I have also identified children who had at least one parent as an artist
who worked at home and shared art-making all through the child's preschool
years. And in these additional, extremely uncommon cases of rich, dynamic,
interactive conditions for building art skills, each of the children by age five or six
was making and relishing remarkable, complex, sophisticated art.

Chapter 31

# TRICKY MIX EXPLORATIONS WHICH CREATE NEW RETRIEVAL PATHS, NEW SENSORY EXPERIENCE PATHS, AND NEW CONNECTIONS BETWEEN THE UNCONNECTED

or most of the familiar activities that we undertake, we've already stored miles of relevant information in the landscape of our vast brain.

## Revisit the Islands of Your Mind!

A neglected factor in many attempts to teach innovation and creativity, and also in many attempts at therapy and rehabilitation, is to very actively promote and enhance the ways in which we *retrieve and bring into focus* more of that vast information to achieve new connections and definite breakthroughs.

An interesting variation on this theme is multiple-session therapy conducted with some version of cognitive behavioral therapy. Clients who enter therapy with persistent problems of depression or anxiety often have been failing to retrieve relevant past experiences in balanced and accurate fashion. For example, a negative bias may lead to easy retrieval of experiences that were sad, humiliating, or threatening while neglecting occasions where the client was happy, effective, and powerful. Similarly, clients with excessive worry may falsely believe that most of the things that they worry about are actually worth worrying about—that is, that the feared outcomes are frequent in occurrence.

Accordingly, powerful positive effects on clients' self-confidence, emotional regulation, and active strategies may be achieved by revisiting positive past "islands" of experience and integrating those thoroughly. More broadly, all of us will benefit by more often seeking out those islands of experience which have recently been neglected; such significant episodes in our past will range across positive, exciting, proud episodes as well as difficult times of loss or frustration. We need to approach that goal with clear intention, with mindfulness, and playfully try experiments with ourselves on how to reestablish contact and bring the energy and vividness of positive past islands into our present and future.

Another domain worth exploring here is that of creative metaphors. In high-level scientific discovery, the prevalence of surprising but highly effective new breakthrough concepts based on innovative metaphors is widely recognized. But the search for new metaphors or new applications of metaphors already in our possession can be a powerful tricky mix approach for all of us. The more often we conduct such explorations, the more we build up the reservoir of metaphors in our mind as well as enrich our tool kit of techniques for finding new metaphors.

## New Sensory Experience Paths

Some of my own transformative sensory experiences during encounters in the wild with bobcats, eagles, herons, mountain lions, and the like have been discussed in earlier chapters.

Now I'll share some remarkable experiences by the explorer and writer David Abram. The following are quotes from his book *Becoming Animal*:

> "Cross-country skiing along a snow-covered stream in the northern Rockies, I emerged from the woods into a small, frozen marshland and abruptly found myself three ski-lengths away from a mother moose. She had been feeding with her child among

the low willows. The moose looked up, as startled as I; she was facing me head-on, her nostrils flaring, her front legs taut, leaning forward. Her eyes were locked on my body, one ear listening toward me while the other was rotated backward, monitoring the movements of her calf. My senses were on high-alert, yet somehow I wasn't frightened or even worried. I took a deep breath and then found myself offering a single sustained, mellifluous note, a musical call in the middle part of my range, holding its pitch and its volume for as long as I could muster. As my voice died away, I already sensed the other's muscles relaxing ... Within a moment, the moose leaned her head back down and casually began nibbling the willow tips."

The approach which David Abram applied in this and many other instances is remarkably similar to what I have discovered in my own explorations of communication with wild creatures. I have talked back and forth, with great pleasure, with loud-mouth bullfrogs. The central key in all human/animal meetings has been to relax as completely as possible, to send out good intentions and feelings, and show lack of any threat together with the simple desire to know and communicate with the other being.

Another interesting situation of animal contact occurred when Abram was kayaking near some Alaskan islands. A great colony of large Steller sea lions was basking on countless rock ledges:

"They're now plunging into the water in bunches, clusters of them tumbling into the brine and swiftly surfacing and then surging—with their torsos half out of the water and with a holy clamor of guttural bellowing—straight toward me! ... There are not many options, no time to think; my awareness can only look on in bewilderment as my arms fly up over my head and I begin, in the kayak, to dance. More precisely, my upraised, extended arms begin to sway conjointly from one side to the other, with my wrist and my splayed fingers arcing to the right, then to the left ...

"As soon as I begin these contortions, the clamoring sea lions rear back in the water and fall silent as their heads begin swiveling from one side to the other, tracking my hands with their eyes. Astonishing! ... All in perfect unison like a half-submerged chorus line. After a couple minutes I drop my hands down to take up the paddle—but straightaway the sea lions start bellowing and surging forward ... Whenever I even start to lower my hands, the dark-eyed multitude lunges forward—so halting my dance is not an option ... My right arm is giving out. Slowly I bring that arm down while the left keeps up the rhythm. The sea lions, weaving from side to side, are now focused on the single, swaying metronome of my left arm.

"Something in that charged encounter changed me ... It made evident in a way I could no longer ignore that there exists a primary language that we two-leggeds share with other species."

## Common Elements of Tricky Mix Cross-Species Breakthroughs

Abram notes that profound deep, new connections between us and another species changes us, and I concur. Once experienced, we can't go back to a position that humans are forever separate from other creatures. Moreover, arising from these experiences are a lasting sense of stewardship and reverence toward nature.

In tricky mix terms, we can identify many of the converging dynamic conditions for such breakthroughs.

- The human/animal encounter surprises us
- We drop any fear or wariness
- We allow ourselves to "flow" with whatever emerges
- All our senses are heightened
- We observe details never before noted
- We send out feelings of desired communication
- We are unusually sensitive to signals from creatures
- We send out a message of acceptance and delight

- Prior similar experiences feed our positive expectations in this new encounter
- A back and forth, mutually responsive interaction emerges

You may recall the episode in Chapter 27 wherein a solitary pelican and I happily connected and strolled along in mutuality on a sunny California pier. That episode fits precisely with the above dynamic tricky mix pattern.

A broader theme in this book is that when we build a repertoire of known dynamic patterns across many different domains, it prepares us to extrapolate and integrate those patterns creatively in fresh encounters of all sorts. For the present discussion, note that tricky mix cross-species breakthroughs change our pattern repertoire and our awareness. A result may be that for important new episodes of purely human interaction we actively seek to bring together more of the cross-species elements, resulting in greater intensity and richness of interactions plus the emergence of new creative breakthroughs.

## New Connections Between the Unconnected

Powerful breakthroughs which positively influence billions of children and adults in the future no doubt will rest upon new tricky mixes which bring together strong conditions/partners that were poorly connected in the past.

An encouraging set of partial breakthroughs led to the honor of the US Presidential Medal of Freedom being awarded to Mario Molina in 2013 for his outstanding leadership on issues of atmospheric science and climate change.

The basis for the award was Molina's energy and knack for richly connecting across "islands" of human endeavor—economics, government, citizen action movements, and multiple fields of science documenting global weather events, global warming, human impacts, declining species, as well as data-based models of the future.

His emphasis on new mixes of collaborative efforts continues after his death in 2020 to inspire others to creatively understand the world and to achieve new mixes of behavioral change that may finally reverse many of the adverse consequences for the Earth after decades of pollution and disruption. In the case of the Earth's ozone layer deterioration from CFC (chlorofluorocarbon) gases, he was a crucial contributor to both excellent scientific research and to effective new polices to limit CFCs. The Nobel Prize in Chemistry in 1995 recognized this component of his career.

## Photography Experience as a Path to Proliferating New Connections

An active photographer builds up over time countless new patterns in their heads and in their portfolio. For choosing what to shoot and with what settings, all those previously acquired visual patterns are dynamic influences on what occurs. This I know from shifting the treatment of my work from a "mere hobby" to treating photography as a passionate art form—although still not my main career. In consequence, some of my work has gained recognition in small galleries and shows and in publication within the book *Children, Pelicans, Planets: Bobcat Magic*.

Another important shift is that I have more and more actively brought what I know of visual patterns into creative efforts that at first seemed unrelated. Like countless other individuals, I find that enriched visual pattern sensitivity and pattern memory feed into tricky mixes for planning scientific research, into persuasion/narrative efforts in social activism, into writing, and into collaborative problem solving.

To round out this chapter, I have prepared a series of my photographs for your inspection and reflection. I offer no specific conclusions of my own. Instead, you are hereby invited to peruse and enjoy the complete series. In this first pass, please try to avoid simple labeling or figuring them out too analytically—just soak them up openly.

*Young Fawn Learning to Walk in a Stream.*

Spiraling Dynamics of Interior of Fossil Chambered Nautilus Shell.

With an Appearance Like an Ancient Gold and Lapis Egyptian Piece of Royal Jewelry, This Monarch Chrysalis Encloses a Slow Transformation From Caterpillar to Butterfly.

Staying Stock-Still While Mother Grazes Some Distance Away.

*Two Snapping Turtles Gently Touch Noses in Pennsylvania Farm Pond.*

*Greek God in Motion But a Bit Off-Balance.*

*Buck and Doe White-Tailed Deer Surprised by Photographer and Camera.*

*Complex Dynamic Tricky Mix Maneuver by Goshawk to Capture Dove in Midair.*

*Battle Between Two Pennsylvania Natives —Brown Trout and Northern Water Snake.*

*Mythical Creatures.*

*Great Egret in Liftoff Shows Off Angel Wings.*

*Dolphin Racing Alongside Our Fast-Moving Boat.*

Okay, I hope you enjoyed exploring those images. Next, I ask you to look back again and ask these five questions:

1. What other events, experiences, or patterns are you thinking about as a result of viewing these?

2. Do you see interesting connections between two or more photos which you find surprising or inspiring? Are sub-patterns similar across different photos?

3. Think in terms of metaphors and symbols. Which of these photos would be a good symbol, meme, and/or metaphor for a project, a goal, an emotional state, a company, or some other entity?

4. Breakthroughs of many sorts have been launched by inventing a new metaphor/symbol or by applying familiar metaphors in new ways. So, would you now have any innovative ideas along these lines?

5. Which of the photographs above might fit with these potential themes/metaphors?

Chapter 32

# POSITIVE AND NEGATIVE SNOWBALLS IN NATURE'S ECOLOGICAL SYSTEMS

H umans have repeatedly throughout history created crises, devastation, and wastelands in our oceans, forests, wetlands, lakes, rivers, plains, and mountains.

Here we will sample both good and bad news illustrations of how Dynamic Systems effects have been created in mature. Because the same basic processes are involved in cascading, snowball-like effects in either positive or negative directions, we will aim for a message of optimism if we respect dynamic processes and act forcefully when we see the opportunities.

## Restoration Processes

- In Yellowstone Park, the number of wolves and other large predators declines
- This leads to a rapid increase in elk numbers
- The numerous elk eat areas clean of shrubs, trees, and most grass
- These decreases in plant biodiversity lead to fewer small rodents

As a result, an area that had abundant diversity may shift to an ecological wasteland, *but* if a new barrier around the area keeps all elk and deer out, creating major re-mixing of conditions, *then* in even five years' time a positive snowball of recovering small plants, small mammal species, birds with useful waste products, and then shrubs and trees may all be set in motion.

True story—fencing in Yellowstone areas with low biodiversity did produce all these ecological recovery outcomes.

## Stubbornly Pushing Toward Negative Snowballs

Decades of *political distortion* of the facts on energy policies, farming policies, deforestation, transportation policies, and plastic and toxic pollution of oceans have fueled a *toxic negative tricky mix* that creates high and accelerating global warming and global environmental chaos. Trump's administration was callous and reckless and shameless in their policies based upon greed and distortion of the relevant facts. But that is just part of the horrible thread across fifty years of irresponsible political actions by the USA and Russia and many other countries. Recall, for example, that a high-ranking aide in the Bush administration needled a journalist for belonging to "the reality-based community." A respect for facts, the aide suggested, was ultimately for suckers: "We're an empire now, and when we act, we create our own reality."

To successfully combat the combined climate crisis and biodiversity crisis is going to require many new experiments with new patterns of positive tricky mixes. Bold ways are needed for drawing into collaborative efforts a rich swath of diverse thinkers from all professions, walks of life, and cultural identities. The French sociologist and anthropologist Bruno Latour has written two books of high relevance around these issues: *Politics of Nature: How to Bring the Sciences Into Democracy* and *Facing Gaia: Eight Lectures on the New Climatic Regime*. Latour argues that climate change and broad negative impacts on nature are forcing all of us to confront truths that seem hard to reconcile but turn out to be two sides of the same coin: (1) reality exists whether we like it or not; and (2) our attempts to apprehend it and change course are contingent on our social context and our own complex self-concepts. Latour explores ways to think through the seemingly *stuck* positions carved out by political polarization and fake news and to somehow find paths toward creative new tricky mixes of constructive dialogue, reframing, experimental tests of new policies, and self-serving yet altruistic actions.

## The Natural World: Sea Otter Mystery

In undisturbed settings in nature, some complex relationships between animal species, plant species, weather, and many other factors have been recognized for several centuries. Yet, many mysteries still arise. Here is a recent one.

Along the coast of California, a thrilling sight is a sea otter floating along on his back looking marvelously cute. Often the otter will be holding food on its tummy that it has recently captured. For example, the sea otter may hold a sea urchin in one paw and a rock in the other paw. Then the otter smashes the sea urchin with the rock, thereby getting access to the luscious soft meat of the sea urchin inside.

The following mystery arose. In many locations along the Pacific coast, the number of sea otters declined precipitously. Many scientists thought there might be a fairly simple solution to the sea otters' demise, namely some disease that had not yet been identified. Much effort went into trying to confirm this hypothesis but no disease culprit was identified.

Solving the mystery required insight from multiple experts and observers. It also required a full recognition of complex, dynamic, interacting factors across time. One factor was that the number of baby seals available for large predators also had declined. The most relevant large predator for baby seals is the killer whale, or orca. Orcas love baby seals and rely heavily on them in their diet, so it was speculated that perhaps orcas were eating a few more sea otters then they usually would. Leave this idea aside for the moment and consider additional cascading effects that were taking place in the Pacific coast ecosystem.

During the same timeframe as the sea otter decline there was an explosion in the number of sea urchins in multiple spots in California, Alaska, Russia, and other places. After a while, scientists recognized that the decline in sea otters was contributing to less sea urchins being eaten, and thus the survival of more sea urchins. Another cascading effect was that kelp seaweed beds were declining in terms of the richness of the kelp. This happened because the sea urchins nibble away at the hanging kelp ribbons at the ocean floor, usually keeping the kelp from getting out of balance. But too much urchin nibbling, with urchin numbers increasing, harm the kelp balance. Further cascading effects occurred because within the healthy biodiversity that the kelp harbors, myriad species of small fish crustaceans, and other organisms thrive, and those, in turn, serve as food sources for larger fish, octopi, and other creatures.

Investigators gradually began to see that the mystery of the declining sea otters was a mystery of the whole ecological system in crisis. Hungrier killer whales had a large impact on a fairly small overall population of sea otters. A clever biologist was able to calculate that the thirty-five-foot eating machine called an orca could easily devour five sea otters a day, and persist in that eating pattern over the course of the year. You can see, therefore, that in one year that would add up to the loss of a tremendous number of sea otters from just one orca. And beyond that, of course, is the fact that orcas travel in packs. Given all that, it is clear that the orcas, in a short time period, caused the decline in the sea otter population. And once that occurred, then there were inevitable cascading effects on sea urchins, kelp, the countless fish and crustaceans that live in the kelp, and many other aspects of the plant and animal balance within coastal systems in the Pacific Ocean. In other words, a new pattern of eating by orcas altered the delicate ecological dynamic tricky mix and soon triggered a continuing negative snowball for multiple species that proceeded to build at "flowing"/"soaring" levels.

Pacific dolphins, as captured in the image below, are affected by these and other related ecosystem changes off the West Coast of the USA.

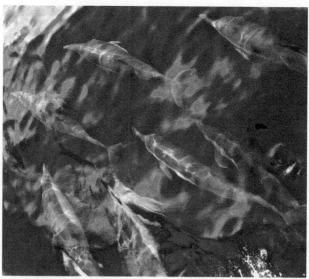

*Remarkably Intelligent and Social Dolphins on the Move.*

In all complex ecological systems, there are opportunities for humans to thoughtfully calculate timely moves that may nudge the systems toward better balance. Such nudges can be towards more biodiversity and toward more healthy conditions overall. That is what we need to keep in mind, as a human species, to preserve and enhance the natural world. But, of course, the same dynamics mean that profit-seeking, reckless actions by human beings—actions that disregard the cascading effects on an ecosystem—may instead lead to rapid and horrifying negative consequences for the ecosystem. We must recognize the complexity of ecosystems fully, identify possible actions that are likely to be influential, and then responsibly choose a set of actions and monitor their dynamic effects on the natural world.

## Ancient Simple Trick with Cascading Positive Effects

Along the coast of Oregon, Washington, and British Columbia, for over 2,000 years the indigenous tribes have harvested and feasted on clams—often with elaborate ceremonies and shared meals.

In the last forty years, though, there have gradually been decreases in the number of clams in the usual clamming spots. Suspected factors include over-harvesting, pollution, and rising ocean temperatures. A combination of such factors cascading dynamically together could mean that little could be done. Humans who love clams and cherish their clamming traditions just might be facing an unchangeable, *stuck* clam decline.

And yet, an ancient practice is being revived: the building of modest "clam gardens." What's involved here is the relatively simple assembly from local rocks of a fence across the sand in the tidal splash zones. The low stone fence arches out across the tidal flat, and when the tide comes in it fills the garden with seawater. Then, when the tide recedes, for a few hours it leaves a shallow pool of water near shore where clams find it ideal to live and eat. In consequence, clams reproduce more successfully and the clams also live longer.

So, by experimenting with an ancient practice, local tribes are able to manipulate one part of the local ecological mix and produce many positive effects. Despite other obvious risks to their species, the clams are resilient enough to take advantage of dynamically changed tidal pool conditions.

Just as we have documented in so many other instances, being *stuck* does not necessarily mean that recovery or major advances are impossible. High capacity for change may still be intact. But actual positive changes may await the introduction of novel re-mixing of conditions that respects the complex, multi-factor dynamics at play. Although the specifics will vary widely, this way of approaching change processes is powerful regardless of the species involved—from clams to rain forest birds and monkeys to people to diverse creatures of the coral reefs.

## Our Laws Concerning Endangered Turtles, Tigers, Whales, and More

The Endangered Species Act (ESA) of 1973 is one of the legal cornerstones of conservation in the United States, and when it works, the results can be spectacular. Bald eagles and peregrine falcons have returned to the skies, wolves and grizzly bears prowl around Yellowstone, and humpback whales ply oceans on both coasts. But deciding whether the ESA is doing its job for other species is much more difficult. There are fewer scientists keeping tabs, less data, and less money. That's why researchers from the Center for Biodiversity looked into the recoveries of marine mammals and sea turtles listed on the ESA: 78% of the populations they investigated saw significant increases after listing, indicating that the law is doing its job.

For the study in the journal *PLOS One*, the researchers looked at the best available data for fourteen marine mammal species like killer whales, fin whales, sea otters, monk seals, sea lions, and five sea turtle species that call US waters home. Because the ESA divides species into distinct populations that it manages individually, the team analyzed twenty-three populations of mammals and eight populations of turtles, finding that eighteen of the mammal groups were on the rise and six of the turtle populations had seen significant gains.

Some of the recoveries are striking. Hawaiian humpback whales, for instance, climbed from just eight hundred animals in 1979 to 10,000 in 2005, which led to a delisting. The eastern Steller sea lion population along the Pacific coast rose from 19,000 in 1990 to almost 60,000 in 2013. And sea otters doubled their numbers to almost 2,700 individuals between 1979 and 2017.

Lead author on this research report is Abel Valdivia at the Center for Biological Diversity in San Francisco. His views fall squarely in line with our emphasis in dynamic tricky mix theory on creating new, multi-component convergent mixes if we are seeking significant positive advances. Over-simplifying is always a danger. Valdivia, for sure, finds no conservation silver bullet that stands out in the data. Each species had an individualized recovery plan and had critical habitat declared, each getting different interventions, whether that was protecting turtle nesting grounds or keeping boats and ships a certain distance from whales. These experimental interventions were monitored and adjusted. Again, there is a great fit with dynamic tricky mix approaches—dynamic monitoring is essential to avoid getting trapped into believing that making a careful and promising plan guarantees success.

Whooping cranes are a case where new dynamic tricky mixes almost certainly prevented extinction. I am very grateful that I have had the pleasure of watching these cranes in flight, their long necks thrust forward and legs trailing behind, their voices echoing loudly. They almost disappeared from North American skies after unregulated hunting and habitat destruction radically reduced their numbers from as many as 1,400 one-hundred-fifty years ago. By the time it was listed as endangered in 1967, the population of America's tallest bird had dropped to just forty-eight wild and six captive birds. In 1978, critical habitat was designated in parts of Idaho, Kansas, Nebraska, Oklahoma, and Texas. Their habitat once stretched from the Arctic coast south to central Mexico, and from Utah east to New Jersey, South Carolina, Georgia, and Florida. They now nest in the wild at only three locations: Wisconsin, Central Florida, Wood Buffalo National Park, and adjacent areas in Canada (a population that winters in Aransas National Wildlife Refuge in Texas). Due to intensive habitat management, nest protection, captive breeding, and reintroductions, the population rose steadily to about five hundred birds in 2006 and six hundred now in 2021.

As noted, I have had the thrill of seeing whooping cranes up close. These days, they often stop in their migration at a preserve near Bloomington, Indiana. Luckily, my brother Craig lives nearby, so my wife, Kathy, and I accompanied Craig for one recent visit there. Craig is a biologist with special expertise in how wild species arise and change in evolution. In addition, he was a key part of that personal tricky mix of inspiring childhood adventures in nature that has fed into my lifelong interests and stewardship actions toward the natural world. Accordingly, watching whooping cranes and other migrating fowl with him was especially sweet.

Political awareness is critical for future successes for endangered species and for overall biodiversity. Despite the kinds of evidence just reviewed on how effectively tricky mix plans can be for multiple species, in the USA, Central America, South America, as well as Asia, right-wing politicians and rogue corporations are frequently undermining ongoing efforts. Thus, all new planning for conservation, biodiversity, rain forest and other critical habitat enhancement, and related goals must re-mix to incorporate not only the best of complex ecological science but also collaborative political commitments across ideological lines and across the globe.

## Renewal of the Land of the Guacamaya

For decades, one biologist from Tennessee spent most of his time over in Honduras, where he loves the Guacamaya or scarlet macaw. These huge, incredibly beautiful birds were endangered despite being the national bird for Honduras. He could see that—to use the language of our tricky mixes—there had been a negative snowball in place for years. Poaching and smuggling/exporting of macaws accelerated, breeding populations spiraled downward, and no effective restraints were placed on these trends. Dynamic tricky mix processes were working in powerful fashion but in horrifying directions.

Lloyd Davidson rose to the challenge of reversing the decline of macaws and took the initial steps to create a modest snowballing positive effect. With just a few collaborators, in 2001 he rescued captive macaws and put ninety of them on a plane and flew them close to land he owned. After accumulating enough macaws to make his private park a tourist attraction, he opened it up and spread knowledge and admiration while generating limited income. The macaws were fed and protected, and so they were tame rather than wild.

Moving beyond this start, breakthroughs of major significance then evolved. A new tricky mix was introduced which led, after just a few years, to a *flowing* mode of progress toward boosting the wild macaw population:

- An idea of a workable plan for macaw release into wild areas overcame prior doubts
- Out of the blue, an experienced leader of a worldwide parrot organization volunteered to help Davidson
- More funds were generated by multiple organizations
- Some funds were used to work with schools to inspire children and parents about macaws and to reduce illegal poaching and smuggling
- Initial macaw releases into wild areas used a transition period where food and contact with humans was maintained as in Davidson's preserve

Then, the dynamic, sustained convergence of this tricky mix led to many successful wild releases and to an accelerating rate of releases year by year. A *flowing* mode of rapid positive change had been reached which excited Davidson's partners all over Honduras.

Next came an even better period where additional factors were added to the mix while maintaining the key factors just reviewed.

A *soaring* level of fantastic progress ensued when the following conditions dynamically joined the Mix:

- Fully released macaws began reproducing
- Each new group of macaws released raised human expectations that a strong future for macaws could be sustained
- Successes led to participation and interest by countries neighboring Honduras
- Davidson's preserve grew in reputation, visitors, and funds generated
- Local citizens increased every year in their help, appreciation, and enthusiasm

So for these beautiful soaring Guacamayas, the fantastic news is that in 2020 there are strong forces for continuing species health and for sustaining the *soaring* transformational positive dynamic tricky mixes.

*Scarlet Macaws.*

# Borrowing Nature's Patterns to Inspire Dynamic Re-Mixing for Human Social Projects

We argued several times earlier in the book that learning some of the complexities of Dynamic Systems in nature is an excellent route to inspiring new, effective strategy mixes for a wide range of projects and issues.

Two must-read authors on such approaches are Richard Louv and Thomas L. Friedman.

Louv, a journalist by training, founded the thriving Children & Nature Network and wrote two influential books—*Last Child in the Woods: Saving Our Children from Nature-Deficit Disorder* and *Vitamin N: The Essential Guide to a Nature-Rich Life*. All of this work points in two directions. First, increasing our experiences and our children's experiences in nature are essential to understanding and relishing nature and to increased stewardship efforts toward biodiversity and sustainability. Second, by doing so we are changed for the better and are certain to discover new and creative dynamic patterns to explore and test in every domain. These include, for example, the domains of conflict resolution, economic prosperity, reduced toxicity, lively and effective education, supporting the arts, and affordable health care.

Similarly, Sim Van der Ryn and Stuart Cowan in their book *Ecological Design* stress geometric patterns, the dynamics of particular places and contexts, and the potential of every person as an influential designer.

Friedman is a three-time Pulitzer Prize winner. He makes important arguments and covers crucial nuances in his paradoxically entitled book, *Thank You for Being Late: An Optimist's Guide to Thriving in the Age of Accelerations* (giving me time to be reflective). I wholeheartedly agree that for most human endeavors, as Friedman argues, we can learn much both by observing the complexity of processes in nature and by finding pathways for resilience.

Antoni Gaudi very explicitly based many of his designs for parks, furniture, houses, and cathedrals on patterns found in nature. His architectural and other design work in and around Barcelona is highly creative and influential around the world. The nature patterns he adopted and re-mixed include visual patterns, underlying structural-functional properties, and elements of joy and inspiration.

The first image below is a nature-inspired roof sculpture. Also pictured below is one dynamically complex set of patterns within the basilica he created in Barcelona, the Sagrida Familia. Note how the multiple pattern layers play out in tricky-mix fashion—the windows and walls themselves, the light as it streams through the windows, and the changing light as the sun's angle shifts and the mix of colors and shapes splash across the interior space.

*Gaudi Roof Sculpture in Barcelona.*

*Light Streaming within Gaudi's Sagrida Familia.*

Earlier, in the chapter on da Vinci's ancestors, we discussed how patterns collected from nature were crucial in the very evolution of symbolic capacities and symbolic cultures over 400,000 years ago. Below is one beautiful, small sculpture from Ecuador utilizing resins from the mopa mopa trees with a complex tricky mix of steps within a technique that is more than 1,000 years old. This colorful sculpted creature bears striking resemblances to many of the animal-inspired designs Gaudi created in parks and the interiors of houses.

*Nature-Inspired Sculpture*
*Created by Quillacingas, Natives of Ecuador.*

Also of high relevance here are sculptures and installation pieces by the Scottish artist Andy Goldsworthy. He deeply understands complex patterns in nature and works with nature as a o-creator of his art.

*Goldsworthy "Welded" Pieces of*
*Ice to Form This Curving Ice Spiral*
*Around a Tree.*

## Chapter 33

# MORE TRICKY-MIX MOVES IN THE FACE OF HIGH UNCERTAINTY AND HIGH COMPLEXITY

Throughout this book, we have stressed the dangers of oversimplified actions and strategies in the face of dynamic complexities. Instead, we have seen illustrated how experimental and imaginative creations of new tricky mixes leads to important, breakthrough advances in all kinds of areas or domains.

Now we will delve further into the nature of certain highly complex, dynamic patterns and the kinds of dynamic tricky mixes that will be appropriate and promising.

## Daily Choices of General Health-Related Actions: Scenario with Fifty-Two Specific Options

Most of us give some careful attention to specific actions that we believe may feed into our optimal health and cognitive performance. We face a very complex and large landscape of possibilities to choose from. These include particular foods and supplements, exercise/workout options, meditative and yoga activities, social meetings, and medications.

But how do we monitor and judge which of our choices are really having the desired positive effects? Complicating such monitoring are two inevitable aspects of the complex dynamics: (1) some actions will show their effects after an uncertain lag time of one or more days, and (2) many actions will contribute far more strongly to progress if they are combined on the same day with other converging actions, creating synergistic and multiplicative positive dynamics (revisit Chapter 11).

Given the above complexities and complications, it turns out that one especially pertinent variety of mixes are "banquet" tricky mixes. Rather than leaving our choices for the day (or week, month, etc.) to intuition, in many instances it is workable to create banquets of actions which work against some otherwise discouraging odds. To understand this, looking at some mathematical aspects is essential.

For the present, consider a set of fifty-two action possibilities of all sorts. For example, consuming vitamin E, lifting weights, jogging, meeting a best friend, consuming a neurotransmitter precursor, using a popular medication, eating broccoli, or meditating. Most of us could generate—please generate your own list on a sheet of paper—more like one-hundred-plus possible choices we have actually taken or have considered taking in the areas of health and cognition.

But whatever choices we make today will interact dynamically with our current state. Assume that only a few choices will groove well with our current state and lead to personal progress. In fact, assume for this discussion that a combination of just three of those fifty-two possibilities will be *optimal* choices for today, and that these three will work together dynamically. Also assume that we are very poor at predicting which three should be chosen today for best progress, yet we go ahead and make three choices. Here's where the probabilities of success can be estimated. There are lots of combinations of three elements out of that total set of fifty-two. Namely, there are 22,100 such groups of three! By random choice of just three actions/conditions, our chance of picking the optimal three is just 1 in 22,100. That's a very, very low 0.000045% chance of success for one day!

A banquet tricky mix provides a spectacular and highly encouraging contrast. If we provide a rich banquet of actions/conditions today, then our current systems—physiological and psychological—will choose just what they most need as a combination of three supportive conditions. In addition, among the other forty-nine elements of the banquet, there will likely still be facilitative conditions, even though they are not the most important for today. Even further, if day after day our systems are dynamically taking advantage of the optimal, most-needed conditions, then positive snowball effects will arise across time as one advance builds the foundation for further advances on later days.

If we can set up the banquet a just described, then it is *certain* that the three most optimal conditions will occur rather than the extremely low chance of 0.000045%. Moreover, we should recognize that this approach also bypasses the "lag" dilemma where three conditions may be available by chance today when they are not optimal but those same three conditions would be optimal if they were to occur four days from now.

You might be thinking of "yes, buts." For example, "Yes, but you could never put together fifty-two actions on the same day." Well, I maintain that there truly are many practical sets of exercise, social activities, food, supplements, and other choices that will fit in the time frame and budget of a single day. OK, not every set of fifty-two promising conditions you think about will be that practical, but, again, many sets of fifty-two will be very doable. Those are the ones to try out and monitor to determine whether the predicted snowballing advances in well-being are created for you.

Please note that if we are implementing this plan—aiming for the complete banquet of fifty-two each day—good things will result even when we show some imperfections. That is, if we average completing between forty-five and a perfect fifty-two, the odds are still very good indeed that on most days we will get the optimal three conditions for that day.

## Reducing Our Pandemic Virus Risks: Scenario with Twenty-Four Specific Options

Experiences across the last nine months of the COVID-19 pandemic for different countries and communities demonstrate that many specific actions have helped reduce risks of infection. Some such actions can be taken by any individual, but other actions require the planning and cooperation of multiple persons in a community.

As in the prior scenario, we will assume that for an individual person on a particular day there are three specific actions that will have the greatest impact on reducing their risk of contracting the virus. In the present scenario, assume that there just twenty-four specific actions under consideration. Among these would be avoiding all indoor spaces except our home, always wearing a mask outside the home, using hand sanitizer after receiving a package, temperature checks, taking supplements that raise our general immunity, taking a test on presence of the virus, trying to persuade health professionals to wear protective body suits and other protective devices, trying to persuade everyone to wear masks when outside the home, avoiding gatherings of more than eight people, making disinfection materials free and wait-free in public building entrances/exits, and trying to persuade officials to do contact-tracing and consequent isolation periods for those with definite exposure encounters. Some of these actions will be difficult, and in public discourse there will be differences of opinion on which choices matter and on whether the cost/inconvenience of an action may be too much.

If we lazy and randomly choose the three actions we like and are able to execute on a particular day, what are the chances that the three which matter most to us for that day have been chosen?

There are many combinations of three elements out of that total set of twenty-four. Namely, there are 2,024 such groups of three! By random choice of just three actions/conditions, our chance of picking the optimal three is just 1 in 2,024, just a 0.00049% chance. Although we have acted with good intentions, the probability of much COVID risk-reduction is *very low*.

A banquet tricky mix, once again, provides a spectacular and highly encouraging contrast. If we provide a rich banquet of actions/conditions today, then our current biological systems will choose just what they most need as a combination of three supportive conditions. In addition, among the other twenty-one elements there will likely still be facilitative conditions even though they are not the most important for today. Even further, if day after day our systems are dynamically taking advantage of the optimal, most-needed conditions, then positive snowball effects will arise across time as one advance builds the foundation for further advances on later days.

If we can set up the tricky dynamic banquet of twenty-four choices daily, as described, then it becomes *certain* that three optimal conditions will occur. Moreover, we should recognize that this approach again bypasses the "lag" dilemma where three conditions may be available by chance today when they are not optimal but would be optimal if they were to occur perhaps four days from now.

## A Note on Basing Banquets on Dynamic Systems for the Planned Contexts

It is important to distinguish the strategies in the above banquet scenarios from a crude "kitchen sink" or "shotgun" approach in which everything anyone has ever proposed is thrown at a problem or issue.

Instead, tricky mix banquets should follow the earlier-noted maxims on approaching action decisions with a model of what we know so far about the factors at play and their interactive dynamics. That model will have different particulars for hospitals, schools, management of ocean resources, or any other variation on the main context(s) under consideration.

Pulling examples from 2020 USA political dramas, an experimental banquet of actions to combat COVID-19 should not include injections of bleach, injection of (totally irrelevant) vaccines for polio or mumps, or borrowing demons from dreams.

## Another Tricky Mix Tactic: Create a Crucial Outcome with a Large Banquet of Dynamic Systems-Based Factors and then Experiment Toward a Smaller, Powerful Banquet

For this scenario, consider sixth graders whose schools have not been able to teach them even a first-grade level of skill in literacy or math.

For these kids, the crucial outcomes sought are any significant gains in math, reading, or writing. By tailoring a large dynamic tricky mix banquet to each child, when highly successful rates of learning do emerge, the next strategic step is to "tinker" or experiment with the banquet. That experimentation might reveal that only one-third or two-thirds of the banquet actions are needed—thus saving some considerable time and monetary costs. Alternatively, even though excellent learning rates have emerged for a child, more variation of banquets offered could lead to even stronger engagement and learning levels.

Regardless, once we see that a child does display excellent learning capacity under rich banquet conditions, we have the ethical obligation to respect that *demonstrated competence* and continue to arrange teaching/learning opportunities that dynamically work in the future. Note also that, for the most part, these children were failing under prior teaching regimes heavily dominated by the flawed viewpoint of "choose the best teaching plan (as believed at the time) and leave other plan explorations behind."

## Improving Narrative Packaging of New Tricky Mixes

Imagine situations in which creative and well-informed teams have come up with innovative mixes with superb chances of achieving project goals if relevant persons are persuaded to perform actions that are dynamically central to progress. Here we address the option of framing narratives about the project as somewhat of a tricky mix of its own.

If a group project has been ongoing for a few years, it will definitely be operating with some dynamic mixes of conditions whether or not leaders or participants are monitoring how the dynamics are playing out.

Consider an international project with the central goals of slowing deforestation of rain forests worldwide and preserving the rain forest habitats, their diverse plants and animals, and their sequestering of carbon for many future generations. Narratives will already be in place for why these goals are important and what actions are needed and by whom. Part of those existing narratives is that the scale and pace of deforestation are shocking—in 2019, tropical regions lost almost twelve million acres of forest.

What if creative teams generated new narratives as distinct mixes that have the potential of being highly positive catalysts? Even if the new set of conditions in the main project does not add or subtract conditions, with the new narratives, new levels and intensities of dynamic convergence may be triggered. For example, teams within the project may shift into more frequent "soaring" or "flowing" levels.

With modern media networks and other modern technologies, it is wise to actively experiment with multiple narrative versions. At the same time, each narrative should try to include dynamic elements that converge to raise the emotional engagement, the vividness and memorability, the ties between the perceived source/narrator and the intended audience, and instances of surprise and humor.

For this kind of international project, new narratives that connect with the life experiences and passions of a new range of individual could be especially powerful. We have already seen earlier in the book how beautiful but endangered macaws became—because of new narrative framings—a new focus of many individuals previously lacking any experience with macaws. Catalytic effects could easily flow from the entry of about two hundred new people. Here is a group of conditions which might emerge with high convergence, high intensity, and the triggering of outstanding progress toward project goals:

- Persons already on the project raise expectations
- Persons already on the project increase the number of actions they take
- Persons already on the project increase the intensity of their efforts and their emotional commitment
- Persons already on the project broaden their outreach communications

Persons already on the project raise more funds

Persons already on the project and newcomers experience highly satisfying *flowing* and *soaring* events

Monitoring reveals tangible steps forward in slowed or even reversed deforestation

Monitoring reveals tangible steps forward in increased biodiversity of myriad species

All these positive changes are also folded into updated and encouraging narrative variations

In Costa Rica, it appears that new narratives have indeed succeeded in drawing lots of small farm operators into active participation in reforestation and active roles as communicators to tourists and locals alike on the satisfactions, sensory experiences, and importance of biodiversity and reforestation.

The government of Costa Rica over the last twenty years has not only created vivid and effective narratives for reforestation, but combined those with land donations and financial and other incentives to small farmers to raise their incomes as well as preserve and enhance cloud forests. The government projects are financed predominantly by a tax on fossil fuels, and they have paid out a total of $500 million to landowners. These projects have saved more than one million hectares of forest, which amounts to a fifth of the country's total area, and planted over seven million trees. The story of one farmer gives the flavor of how a complex mix of activities can be successfully facilitated by a complex but effectively communicated tricky mix government plan.

Pedro Garcia includes in his stewardship mix the mountain almond—a tree which can grow up to two hundred feet tall and is a favored nesting spot for the endangered great green macaw. Garcia has worked on his seven-hectare plot in northeast Costa Rica's Sarapiquí region for thirty-six years. The plot has been transformed from bare cattle pasture to a densely forested haven for wildlife where the scent of vanilla wafts through the air and hummingbirds buzz between tropical fruit trees.

Garcia has restored the forest—home to hundreds of species from sloths to strawberry poison-dart frogs—while also cultivating small crops from pepper vines to organic pineapple. To make ends meet, Garcia relies on ecotourism—he guides visiting biologists and ecologists around the plot for a small fee—and modest payments from the government programs.

## Making Values Both Salient and Intense

The likelihood of breakthrough successes in projects will also be enhanced by explicitly bringing in higher values. That is, rather than stressing just the goals of a project and planned steps toward that goal, invocation of high values can be mixed together with other conditions in dynamic, convergent fashion. These messages on value serve as catalysts to high emotional engagement, persistent effort, and a sense of personal efficacy.

Take, again, the case of international enhancement of rain forests and the biodiversity therein. Beyond the most central goals, ties to many high values suggest themselves:

- Reduced inequality across cultures and communities
- Sacredness of nature
- Increased understanding and connection to the natural world
- Obligations to future generations; altruism rather than personal profit
- Spiritual well-being

- Enhanced socialization of children into stewardship roles regarding natural places and all sentient beings

## You Were Expecting A-B-C-D but Instead Got Outcomes L-M-X-Y, So What Are Your Next Choices and Actions?

You and everyone else must choose from the options which are *currently in each person's awareness*. However, since we were not good at predicting outcomes so far, it may be an ideal time to try to expand our awareness and our options. Dynamic tricky mix thinking should help.

Let's approach this through the image below and some metaphoric extensions which flow from it.

*Swan with Cygnets—Decisions on How to Best Protect the Young in Constant Dynamic Flux.*

Now imagine that this scene represents, in order, a series of contrasting situations:

- Educating young children in your family and/or your community
- The small company you just launched and your employees
- Persuading others to take public health actions which make your children safer

In each of these contexts, describe what you believe to be components of a positive dynamic mix already in place. Then identify probable threats to the success of that mix. Further, propose additional dynamic tricky mix components which may be powerful agents to counter such threats.

## Chapter 34

# LIVING OUR LIVES AS TRICKY MIX EXPERIMENTERS: REVIEW, PART ONE

H ello again, reader. Regardless of whether I have ever met you personally, I know two key things about you. First, you are wonderfully unique, and on a unique life path. This is true, as well, for your children, your spouse, and close friends—everyone you know.

> As the famed poet Mary Oliver says in one poem:
> "I think of each life as a flower, as common
> as a field daisy, and as singular,
>
> . . . . . . . . . . . . . . . . . . . . . . . . . . . . . . . . . . .
>
> and each body a lion of courage, and something
> precious to the earth.

The second key thing I know about you, and everyone, is that by bringing to bear an active awareness of dynamic complexities and striving to create in your life new explorations along the lines of new dynamic tricky mixes, you will increase the frequency in which you discover new, satisfying experiences. This approach will facilitate for you many breakthroughs to new levels for all manner of goals, dreams, skills, and projects.

Most of us at times are eager for simple solutions. We say, in effect, "Give me the tried-and-true approach," and, "You experts work it out for me." Then we commit to that approach and don't look back until some catastrophe interrupts our pathway. Oftentimes we ignore the biases that lie behind someone else making money and/or gaining power out of our choosing their "guaranteed" solution.

Instead, we have seen remarkable successes arising out of a very different overall strategy that assumes our individual differences are profound and interesting. Because we are unique and dynamic, if we want high levels of success and satisfaction, we need to embrace active experimentation to discover what may work for *us*. Further, because many complex factors are dynamically at play for any project or goal we pursue, it is essential for us to try and monitor multiple tricky mix variations to give us feedback on which pathways are outperforming their "competition."

Frequently, the outcomes are surprisingly good even though the particular new mix of conditions persons or organizations create did not require expensive, complex new resources. Because the tricky mixes we experimentally introduce are informed by the complex conditions we are encountering, our new pathway can be powerful without itself being complex or difficult. As Warren Buffet notes (and as we highlighted earlier) for business and investment contexts: "It is not necessary to do extraordinary things to get extraordinary results."

## Powerful Moves You Can Employ to Facilitate Breakthroughs for Any Project or Goal

Below we will cover many possibilities without listing out an exhaustive set.

You will find that it is helpful to think about three kinds of innovative moves available to you when facilitating breakthroughs for a project or goal:

1. New paths to fuller analyses of complexities
2. New paths to actions and experiments
3. New paths to valid and sensitive assessments and monitoring

## Be Emboldened to Act If You Have a New, Outside-the-Box Analysis of Complexities and Possibilities

Although naysayers may strenuously attack your new perceptions as too wild to be useful, those attacks carry energy which you can rechannel and mix into your projects.

Consider the issue of the huge amounts of plastics accumulating in the oceans and disrupting plant and animal life there while also contributing to global warming. This problem has been accelerating for decades. Further, there are so many countries, industries, and practices contributing that slowing or reversing such plastic accumulation has seemed hopeless to countless commentators.

A new organization, Common Seas, brought fresh perceptions and fresh calls to action on this issue. They began with regional efforts and then expanded internationally. Totally in accord with *Dynamic Tricky Mix* thinking, they actively explore, in parallel, multiple bold strategies for reducing oceanic plastics. Common Seas, for example, were a key collaborator in the 2020 landmark report, *Breaking the Plastic Wave*.

## Confront Inconsistent, Contradictory Beliefs Evident in Your Words and Actions: Launch New Explorations and Experiments

If you would never expect to see how much music potential your children have without immersing them in significant music-learning opportunities, why would you assume that your children's art potential is minimal without first experimenting with learning time for your child with talented artists?

If you massively research social media evaluations of new products such as air fryers or cell phones, why would you passively accept your doctor's advice to try a new medication? Why not get more information, and if your new information indicates some promising alternatives delay trying the medication until you have experimented with multiple non-drug approaches? And in such experiments, why not be sure to share your ongoing results with friends and others to see and compare whether there are others who are finding similar outcomes? Recognizing dynamic complexity carries with it essential relief from any pressured feeling that you must figure it all out by yourself and from any bias that there must be some simple solution.

## Learn from "Surprising" and "Tricky" Paths to Breakthroughs in Businesses and Other Organizations

Guy Raz for many years has hosted the media show *How I Built This* and now, in 2020, has released a book by the same name.

Again and again in his accounts of how certain entrepreneurs achieved breakthrough successes, Raz emphasizes how tricky and unexpected their paths to success proved to be.

Consider the following inspiring example. Among the panoply of "energy bars," Clif Bars have proven to be a perennial favorite. Their arrival on the market began with a gleam in Gary Erickson's eyes and his Cuisinart that had to be ready for tons of work.

Erickson kept mixing and re-mixing bar variations, always seeking feedback on each. The process proved frustrating—taking hundreds of attempts. Finally, after seemingly endless "failures" he found a bar variation which was truly satisfying and exciting, and this bar was the first to actually go to market.

Another powerful set of breakthroughs in the area of food possibilities was led by the passionate chef José Andrés. He came from Spain and brought into US restaurants the concept of gourmet small plates—multiple plated dishes forming a full meal. His restaurant, Jaleo, in Washington, D.C., was an outstanding success and fostered many spinoffs. In the next section we highlight his benevolent activities.

## Your Passion can Drive New Intensity Levels on Converging Components of Dynamic Tricky Mixes—and High Intensities Create Dramatic Effectiveness

Chef Andrés was the founder of an organization devoted to feeding the hungry in a wide range of crisis situations, from floods, earthquakes, and extreme poverty to the COVID-19 pandemic. Through his own passion he has harnessed the passion of countless other chefs, kitchens, and volunteers to implement hunger relief in creative ways. We saw earlier that, in theory, an increase in the intensity level of a component in a tricky ix may dramatically increase the probability of breakthroughs occurring. For World Central Kitchen, the organization under discussion here, those intensity increases were generated for many important component factors. These include higher quality of relief food, timely responses to crises, large-scale deliveries where crises are extreme, repeat volunteers, and clear and compelling narratives of compassion and shared effort.

As of a recent review, World Central Kitchen has been responsible for the delivery of over thirty million meals. Those meals have inspired hope and persistence and, in many cases, saved the health or even the lives of the recipients.

## Escape the "Tyranny of the Average" Views of Science and Public Policy

We have repeatedly seen major public policy positions and major approaches to scientific description too heavily rooted in finding the typical or average or "mean" levels of persons or situations. This is far too simplistic.

Dynamic Systems tell us that what will change one person or organization positively may be dramatically different than what works well with another person or organization. When we look toward, and spend huge dollars on, an "average" target, we are likely to miss most of the targets because they are so wildly diverse. Concentrating resources and hopes on one simplified target is arrogant, wasteful of monetary and human resources, and a hindrance to identifying what works for different individuals/groups/organizations under different contextual conditions.

## Recognize that Similar Dynamic Processes Underlie Our Peak Performances and Our Worst Performances

Here we will first revisit two dynamic puzzles.

In Chapter 5, these were Cases 4 and 5: The King of England and the Male Superstar of India's Bollywood Movies.

Each of these guys needed a breakthrough that would allow them to overcome a malady that persisted from about age four right into adulthood and crucial roles.

So, reader. What was the shared malady, and what were the dynamic mix routes to becoming, basically, *unstuck* from the malady?

To solve this puzzle, we will take a deep Dynamic Systems dive not only into Prince Albert/King George's dilemma, but stutterers everywhere and their peak highly fluent episodes of speech along with why they experience non-fluent valleys. See how this approach compares with whatever ideas you proposed.

In the movie, *The King's Speech*, we see dramatic portrayal of Prince Albert's desire to give a crucial and fluent speech to the English public. He wanted a peak performance but was acutely aware of his tendency to stammer (stutter). Earlier in Chapter 5 you had the chance to predict how he found a dynamic tricky mix that helped him before critical speeches to his nation and other high-pressure events.

Well, here's the story. Prince Albert (before becoming king) was able to work with the most famous therapist on these matters in all of England. A key component of a new mix was the prince's confidence, which his therapist was able to instill because of his prior success with many clients.

Now a brief digression. It turns out there is a remarkable backstory to how the public narrative in the movie was created. We may observe that another breakthrough occurred, one that led, after twenty-five years of delay, to the breakthrough narrative of *The King's Speech* winning four Oscars in 2011 and 2012. It won Best Picture, Best Actor, Best Director, and for David Seidler, Best Original Screenplay.

Among those listening to the actual wartime speeches was a seven-year-old British boy who, like the king, had a wealth of words but could not get them out. "I was a profound stutterer. I started stuttering just before my third birthday. I didn't rid myself of it until I was sixteen. But my parents would encourage me to listen to the king's speeches during the war, and I thought, 'Wow, if he can do that, there is hope for me.' So he became my childhood hero."

This is what David Seidler, who wrote the movie, told journalist Scott Pelley of the CBS show *60 Minutes*. Seidler didn't want to tell the tale in a movie screenplay until he secured permission from the late king's widow, known as The Queen Mother, who consented with the proviso that nothing would appear while she was alive.

In 2002, Seidler was freed by her death to go to work. He found the theme of the story in the clash between His Royal Highness and an Australian commoner who became the king's salvation, a speech therapist named Lionel Logue. His grandson, Mark Logue, discovered a trove of documents in an attic. These included the original copy of "the speech" typed out on Buckingham Palace stationery.

"What are all of these marks? All these vertical lines? What do they mean?" Pelley asked, looking over the documents.

"They're deliberate pauses so that the king would be able to sort of attack the next word without hesitation," Logue said. "He's replacing some words; he's crossing them out and suggesting another word that the king would find easier to pronounce."

"Here's a line that he's changed: 'We've tried to find a peaceful way out of the differences between my government.' He's changed that from, 'my government' to 'the differences between ourselves and those who would be our enemies.'"

## The Complex Tricky Mix that Supported the King

For Prince Albert—the soon-to-be king—all the components just discussed were combined to create the following convergent dynamic tricky mix.

- Confidence is high
- Most practice is under relaxed, easy-fluency conditions
- Cues are used to practice under harder conditions
- Success in practice leads to even higher confidence
- During actual prime-time speeches, useful cues and reminders are present
- During actual prime-time speeches, his therapist and wife are also present to give emotional support
- The written copy of his speech is heavily marked up and edited to cue breathing pauses and to eliminate any "hard" words
- He is persuaded to use the trick of standing on a table while speaking

During a speech, all of these conditions would interact and dynamically feed into positive emotion, focus, narrative force, and confidence.

## Actor Hrithik Roshan

Both similarities and differences to the king arose in the modern case of the famous Indian actor Hrithik Roshan. Perhaps the most prominent difference is that Hrithik from an early age sought the limelight and craved public appearances whereas King George VI hated the limelight and appeared only out of a sense of duty. As Colin Firth, the actor who portrayed the king, summed it up:

"It's a perfect storm of catastrophic misfortunes for a man who does not want the limelight, who does not want to be heard publicly, who does not want to expose this humiliating impediment that he's spent his life battling."

Hrithik Roshan, instead, kept seeking the spotlight from an early age and persisted in the enormous efforts put into creating adequate tricky mixes that worked for his achieving fluency.

As an adult actor, he has always spoken openly of his stammering and his speech therapy. His openness and his charitable contributions have done much to help others in India deal with issues of stuttering.

One of his many movies, the 2006 blockbuster *Krrish* saw him play a super-hero. Since then, Hrithik Roshan's career has been spectacular.

In 2009, Roshan spoke very openly about his stuttering on the popular show *Tere Mere Beach Mein*. On the show, he spoke about how he was teased as a kid and yet managed to fulfill his dream of becoming an actor. He said:

"Everything seems normal until you start talking; you get *stuck* and you don't know why. Right from your toes to the ends of your hair strands, you are in complete shock. Your heart palpitates, you don't understand, and you are aware of people looking at you. You can compare it to hell."

Roshan continues to work diligently to maintain a mix of strategies to support his successes, including about an hour of speech therapy exercises most days. Here he describes the occasion of accepting his first acting award at what is the Indian equivalent of the Oscars:

"I was getting the Best Debut Actor for my first movie in Dubai, and I wanted to say 'I love you Dubai' in my speech at the award function. But I couldn't say 'Dubai.' I wanted to practice saying 'Dubai,' but for that I had to scream 'Dubai' loudly. I couldn't do that in my hotel room because my voice was being heard outside. I went to the bathroom and yelled 'Dubai,' but even then it could be heard by people outside. Thankfully, there was a big cupboard in the room. I locked myself up in the cupboard and practiced saying 'Dubai.' Finally, at the awards function I was able to say that with a flow."

That incident vividly captures on-the-spot dynamic tricky re-mixing.

## Finding Stuttering Where It Is Not Expected

A simplistic view of stuttering is that when an idea is formed and motor commands for speaking are assembled, then something goes wrong in the actual execution "locally" of those commands. Dynamic Systems and modern understanding of the human brain say otherwise.

Expressive communication sometimes breaks down even when there are no motor commands for speech at play. Pause, reader, and think on how that might take place.

The first and best-documented examples come from observations of stuttering within sign language.

Stammering or stuttering behaviors that show dysfluency include hesitations/blocks, repetitions of words in sign language, repetitions of first syllables in finger-spelled sequences, loss of fluidity in a sign, and keeping an original handshape while repeating its movement.

Very surprising, huh? Individuals are getting *stuck* in their expressive performance, but it's not because lips, tongue, larynx, and breath are just not cooperating in the moment. More complex dynamics appear to be involved.

But wait, there's more along these lines. For example, the case of a trumpet player who stutters when a score requires sequences that are too rapid. His own description of how that compares to his occasional stuttering in speech is revealing:

"As one who has stuttered, and still does every now and then, I can tell you that it feels exactly the same."

## Central Processing Errors are a Common Dynamic Tricky Mix Basis for Stuttering in All Expressive Domains

Given the surprisingly wide variations we have just reviewed, this heading is crucial to understanding what is happening. In all the cases we have considered, the individual has many fluent episodes in which they display high competence in the domain—speech, sign language, playing a musical instrument. Thus, all the relevant skills are already stored in long-term memory. When dysfluencies/stutters occur, it is because, dynamically, multiple factors of negative expectations, anxiety, over-arousal, distraction, and so on disrupt the planning and execution at central processing and executive-function levels.

## How Phantom Limbs and Stuttering Reveal Common Complex Neural and Action Patterns

Our fabulous brains create both highly functional, appropriate actions and dysfunctional actions through basically the same underlying dynamic patterns. Earlier we discussed Ramachandran's work on the strange syndrome of illusory "phantom" pain. Breakthroughs in treatment occur when new strategies avoid simplicities and instead fully recognize the complexity and interrelatedness of ongoing neural activations. Here is one revealing summary by Ramachandran:

"The idea that the brain behaves like a computer, with each module performing a highly specialized job and sending its output to the next module, is widely believed . . . This is not how the brain works. Its connections are extraordinarily labile and dynamic. Perceptions emerge as a result of reverberations of signals between different levels of the sensory hierarchy, indeed, even across different senses. The fact that visual input can eliminate the spasm of a nonexistent arm and then erase the associated memory of pain vividly illustrates how extensive and profound these interactions can be."

There are many related phenomena. These include crossover dynamic sensory events—for example, musical elements evoking vivid visual color experiences, or just the reverse. These are fairly common in the lives of certain individuals, although the events may appear "illusory" for the majority. Sacks, the neurologist, richly analyzes many such phenomena.

## A Reminder on Balanced Strategies When Creating Breakthroughs

If you try to leave everything to chance and serendipity, you will likely experience failures and missed opportunities which could easily have been avoided.

In contrast, if you try to approach everything through planning, you run the risk of persisting in some of your own ego-invested "great plans" long after they have ceased to be effective. You are likely also to miss out on many fresh perspectives and transformative events.

Accordingly, it is wise to aim for *a synergistic, balanced tricky mix* of open explorations incorporating totally new contexts and chance events together with flexible planning which seeks to avoid oversimplification and incorporates extensive monitoring, and adjustment of plans. Also, integrate insights from both divergent and convergent ways of thinking.

# Chapter 35

# LIVING OUR LIVES AS TRICKY MIX EXPERIMENTERS: REVIEW, PART TWO

## Shoot for the Moon! Be Amazing! Know That Truly Remarkable Breakthroughs May be Achieved by Repeated Nudges toward Ongoing Complex Dynamics

N umerous examples of simple nudges triggering important changes in complex systems are scattered throughout this book. In this chapter, we explore several additional episodes.

## When a Serious Problem for a Family, a Community, or a Country is Solved by Only 1% of Those Trying—Take the Optimist's View

Does that sound paradoxical? Well, the pessimist's view might be that the 99% failure rate indicates it is hardly worth trying to solve the problem or issue.

However, there is a strong basis for optimism *if* a dynamic tricky mix perspective is applied. That is, if dealing better with the true complexity of the problem leads to radically new efforts by the 99%, then progress is likely. What is essential is that we put egos aside and learn from what is really going on when the 1%

succeed. Analyzing their strategies, creating new experimental efforts, and monitoring and adjusting those efforts all should be approached from the framework of dynamic tricky mixing.

## Modern Science Updates Darwin's Visit to the Galapagos Islands in the Pacific

Darwin was struck with how these isolated islands led—over long stretches of time, he claimed—to wonderfully unique species. Giant Galapagos tortoises with their great size, and finches whose sole food is nectar from cactuses provide clear examples.

Modern biologists have brought computer technology, modern statistics, and video recording to observations of the same species Darwin identified. In consequence, for some bird species, researchers had amassed exquisite detail on how birds in the species looked and behaved.

Then nature gave an interesting nudge. A weather/wind pattern in the Pacific led to unusually dry, hot weather in the Galapagos. During the resulting drought, food sources for finches declined so abruptly that natural selection kicked in strongly and many of the birds died for lack of food. As detailed in the chapter on snowballs, a cascade of events led to a lightning-fast shift in species characteristics—the surviving birds showed a shift toward new beak characteristics. This demonstrates that, unlike Darwin's beliefs, evolutionary changes can be extremely rapid and can be created in just a single generation or two if there is a shift in the environmental dynamic mix.

## Is it a Boy or a Girl?

In the "land" of freshwater turtles, a simple nudge of water temperature up or down will lead to far more male eggs, far more female eggs, or roughly equal numbers for each gender. So when just one component of the reproductive tricky mix changes, the outcomes are striking.

The physicist Leonard Mlodinow makes the following statement in his book *The Drunkard's Walk: How Randomness Rules Our Lives* that fits these and myriad other examples we have witnessed: "Extraordinary events can happen without extraordinary causes."

## Seek Ways of Merging Seemingly Separate Tricky Mix Streams

To continue rejecting too-simple approaches, we need to ensure that the examples/breakthroughs highlighted be considered stepping stones to even more complex dynamic tricky mixes. If higher-level integrations and intermixing of effects from multiple ongoing tricky mix pathways are created, we can expect enhanced feedback for each contributing mix. Also, it is likely that highly positive, unanticipated outcomes will start to emerge and cascade forward toward positive future dynamic conditions.

Seem pretty abstract? OK, let's consider some specific possibilities.

From Dynamic Puzzle #1 in Chapter 5, we can recognize that if an individual experiences great health benefits through implementing improved probiotic strategies, those benefits include many complex patterns of digestion, production and regulation of neurotransmitters, production and regulation of hormones, immune systems, and regulation of inflammation. Suppose that for several years an ongoing probiotic-initiated dynamic tricky mix along these lines was succeeding as it did for Professor Dietert, who then broadly shared his approach.

Now let's consider what may happen when we merge all that with a separately initiated tricky mix with a different focus.

## Increased Exercise as a Tricky Mix Focus

We will assume that the improved physical well-being from probiotic supplements is a foundation for entering without pain into exploring increasingly intense and varied exercise/activity plans. In turn, these activity changes then cascade into more dopamine and other neurotransmitter availability. This leads, also, to better mood and confidence and even more varied and constructive physical activity. All these then feed back into even better gut health and high-intensity regulatory processes.

Thus, the dynamic integration of effects from a probiotic-initiated dynamic tricky mix and increased exercise as a tricky mix focus led to a dramatic win-win set of interacting and mutually facilitative breakthroughs.

Next, further exciting outcomes are generated when these are combined with yet a third major re-mix.

## Nutritional Changes Specific to Neurotransmitter Enhancements as a Tricky Mix Focus

Unpublished case studies support the conclusion of high benefits from this particular focus. Further research is on the way.

By choosing foods wisely and taking daily supplements aimed at enhancing the very complex neurotransmitter system, new cascades of positive outcomes are generated. Included are cognitive improvements in: attention, memory, learning, creativity, perception, and mood. In addition, there are dynamic changes beyond those generated by the first two tricky mix strands above in complex patterns of digestion, production and regulation of neurotransmitters, production and regulation of hormones, immune systems, and regulation of inflammation.

In combination, all three strands of new tricky mixes work together dynamically and lead to dramatic enhancements in wellbeing. "Soaring" levels of breakthroughs are created and sustained.

## Leapfrog Start-Up Obstacles by Utilizing Assets That Look Worthless to Most

Tino (Constantino) Aucca Chutas is an influential and award-winning pioneer for Peruvian endangered species projects. Among grateful species are the long-whiskered owlet, the yellow-tailed wooly monkey, and the marvelous spatuletail hummingbird.

A key to initial projects—continuing to this day—are plots of land that were severely degraded. They were lacking in both plant and animal diversity and considered of low to zero value by most Peruvians.

Tino's very first step was to secure an unused plot of ground in the mountains of northern Peru which would support the new habitat needed for a hummingbird. The marvelous spatuletail is unique among hummingbirds in that it has only four tail feathers. The tail of the adult male flourishes two great spoon-shaped "spatules" that radiate a metallic, purplish gloss. The males compete for females by whirling their long tails around their bodies in an amazing courtship display.

Tino partnered with the American Bird Conservancy to manage one hundred acres. There they planted 30,000 saplings of native bushes and trees. Soon, a rich tricky mix cascade occurred with better and better habitat including red-flowered lilies and their nectar, insects favored by the hummingbirds, and other flora and fauna. Then the spatuletails returned, nested, raised chicks, and contributed to pollination of flowers and to the overall ecological balance.

Similar efforts on other plots produced similar positive snowballs of endangered species recovery plus enrichment of overall biodiversity. Further, economic benefits have been flowing to the local cooperating communities. These benefits include income from coffee plants and visiting eco-tourists.

## Lessons from Dynamic Puzzle #11—Early Tools Called Hand Axes Piling Up on African Lakeside From 600,000 to 900,000 Years Ago

Let's recap that puzzle from Chapter 5. Given the short life spans for the period, why would huge amounts of time be invested in making thousands of apparently surplus hand axes? And why were these shared in just a few locations around a particular African lake at the Olorgesailie site in Kenya?

Hand axes were stones with some sharp edges, earning their name because most were roughly the size and shape of a human hand and because they could be held in-hand and used as a tool. In the above time frame, it is easy to see that there would be utility in cutting hides and flesh of prey. There would also be utility in becoming increasingly proficient in the manufacture of these tools—those parts of early culture are not puzzling.

But how can we account for hominins meeting and sharing hand axes between clans and an accumulation of surplus hand axes at a number of favored spots near the lake? I believe that, again, we are seeing how reframing and reinterpreting the familiar can be keys to powerful breakthroughs.

We have seen that the collection and sharing for communication of natural objects such as "fish rocks" or "face shells" provided a crucial entrée to symbolic behaviors. It was a "lazy" way of providing new value to known physical objects by reviewing and reinterpreting.

Something similar happened in African lake areas where the abundant hand axes were created. Once it had become routine to locate needed raw stone materials and to share basic "flaking" steps, many new craftsmen would have joined in these enterprises. Moreover, they may have shared natural stone objects which already had edges which rendered them effective tools.

Then, a fundamental quantum shift grew out of meetings between these usually separate clans. Awareness of how different or how similar hand axes were from different hominin craftsmen and their clans was bound to increase exponentially as more and more hand axes were viewed side by side. Such awareness fed into some expansion of flaking strategies. Yet, far more important was that some members of some clans began to reinterpret some hand axes as symbols for absent objects or entities. They found it interesting that some patterns in the details of a hand axe made it seem like a "lion" or "fish" or "antelope head."

When some of these symbolic reinterpretations became shared, they entered into fuller and more precise communication exchanges. Moreover, clan members who focused on these symbols soon began a reframing of the purposes of tool-making to include intentionally planning a tool to have rich symbolic properties.

In consequence, all of these shifts around hand axes and trading of hand axes expanded at the lakeside sites and led over time to a considerable accumulation. The trading between clans was based not only on utility of particular hand axes, but also upon the *symbolic value*. The "best" of such artifacts might have brought high "prices" in varied exchanges.

This account has the advantage that there are many physical objects from prehistory which varied in their symbolic appeal. Unlike many other anthropological or psychological speculations on how cultural behaviors were evolving, the tangible physical evidence from many sites is a strong foundation. Further, as also discussed in earlier chapters, increases in tool sophistication during this time were far too minimal to account for rapid advances in brain size and sophistication. In contrast, rapid cultural adoption of symbolic communication objects would lead to potent differences in survival rates, selection for the better symbol-finders and symbol-makers, and consequent advances in hominin brains and related behaviors flowing therefrom.

Pictured below are varied examples of hand axes and related stone shapes from diverse regions across the time period between 900,000 years ago and 200,000 years ago.

1 cm

*Incorporated Fossil Adds Decorative, Symbolic Value to This Stone Hand Axe.*

*This Ancient Hand Axe Reminds Us That Many Stone Varieties Across Europe, Asia, and Africa Were Transported and Incorporated into Hand Axe Shapes.*

*Top—I Found This Stone That Is Naturally Shaped Like A Classic Hand Axe, Middle—Stone With Sharpening Through Abrasion in Tide Zone Giving Appearance of Hand Axe/Scraper Tool, Bottom—400,000-Year-Old Large Symbolic Artifact from Swanscombe, UK; Size Precluding Actual Use as a Hand Axe Tool.*

## Beware of Anyone's Claim That One Simple Commitment by You Will Bring You a Long Cascade of Positive Outcomes

A showcase example for why this precaution is needed lies in the history of some "miracle" claims for the education of children with autism spectrum disorder. A simplistic procedure of "facilitated communication" was widely adopted, as we saw in the chapters on sacred cows. Yet, that approach was not properly evaluated for a long stretch. In the end, rigorous data debunked the claims that children with no literacy or academic success were suddenly able to learn well and communicate well just as long as an adult "facilitator" provided physical support for the student's typing into a computer.

Wildly positive claims and suggestions of how a single new factor will change your life for the better in countless ways are thrown at us all the time in media commercials by drug companies. At one level, these are simplistic and absurd—we *know* we should ignore the claims. Yet, the drug ads seek to create a devious and effective negative tricky mix where the drug name is linked with positive emotions, high hopes, caring, love, comfort, and "you can trust us." So, the "hook" that may bite us is believing that by simply taking their injection or consuming their pills, all good things will flow. That is, the ads, by their image manipulations, try to persuade us that their products, beyond their main focus, will magically create many extra benefits in our lives—greater wealth and status, and better sex, relationships, strength, and overall well-being.

Similarly, many politicians in the USA and other modern industrial societies push and exploit a myth about taxes, namely the simplistic view that if taxes are kept low, everyone will thrive economically.

Instead, despite low-taxing policies in the USA, rising inequality is one of the biggest social and economic issues of our time. It is linked to slow economic growth and it fosters social discontent and discord. So, given that the five Nordic countries—Denmark, Finland, Iceland, Norway, and Sweden—have some of the world's highest measures of equality, it makes sense to look to them for lessons in how to re-mix social and economic strategies and build a more equal society.

The Nordic countries are all social democratic countries with mixed economies. They are not socialist in the classical sense—they are driven primarily by financial markets rather than by central government plans. These countries demonstrate that there are multiple paths toward a re-mix of strategies that succeeds in sustaining both economic success and high levels of equality.

Given this state of affairs, remarkable breakthroughs socially and economically may be feasible if bold, complexity-respecting dynamic re-mixes are introduced in the USA and other countries with high income inequality and relatively weak economic performance.

## Crucial Dilemmas Are Raised by Changing Landscapes of Educational Achievement and Life-Long Access to Further Enhancements of Skills and Knowledge

Sociologist David Baker in his much-honored book *The Schooled Society* captures the far-reaching global changes in education across the past fifty years. I concur with his emphases on how systematic and dynamically synergistic these changes have proven to be and on the need to squarely face the complexities of modern global education and its many cascading effects on modern culture.

One dilemma these new patterns raise is at the level of personal choice and responsibility. Precisely because so many access points are now available for gaining skills, knowledge, and social connectedness, it becomes more complex and often perplexing to decide on what to choose. From the perspective of the many episodes we've reviewed, there are many clues to wise strategies. For example, simplistic "choose the best path" attempted solutions are less likely to succeed than a more experimental and open approach to achieve productive Dynamic tricky mix pathways. These sort out on a personal level which choices are effective. Rather than trusting intuition or advice from presumed authorities, the exploration of multiple options with full respect for complexity should be approached with an emphasis on actively monitoring and adjusting ongoing strategies.

Similarly, the actual—versus presumed—impact of formal education for twelve plus years on individual and group levels of thinking, emotion, social communication, and cooperative skills and tendencies requires valid and continuing monitoring. Diversity in outcomes will be expected because of the complex mixes of contributing and dynamically interactive conditions, including the unavoidable modeling of behaviors and attitudes by whoever the teachers/coaches/parents/peers are. From the viewpoints presented in this book, society would be wise to promote and assess a set of skills which are by no means guaranteed by most current educational programs. Among these vanguard skills and characteristics are: evaluative and monitoring skills, curiosity and openness to uncertainty, critical thinking, integrative skills, mindfulness, empathy and cooperation, stewardship of nature, questioning and experimentation in the service of discovery, creativity, and skills in bringing humor and joyfulness into both career and family endeavors. Chapter 17 provides diverse examples of how certain dynamic tricky mix approaches may promote such skills and attitudes along with other positive, complex changes in self.

## Your Brain Is a Massive, Complex, Incredibly Fast and Dynamic Instrument for Spotting, Storing, Integrating, and Extrapolating Diverse Patterns—Appreciate This Gift and Make the Best Of It

You have seen so many times in the episodes within this book that remarkably intact and powerful brains leap to take advantage of new, rich tricky mixes—even when a person has been woefully *stuck* for extended periods before radically new opportunities are introduced.

We will briefly revisit what happens when a child with only rudimentary art skills shifts into *flowing or soaring* rates of progress toward significant art skills and profound enjoyment of making art. Please glance at the graphic below showing a potential pattern of interactions between a child and an adult artist.

*The Extremely Rare but Powerful Tricky Mix of Highly Complex Art Challenges Converging with Great Positive Engagement.*

What holds true for a child's art progression holds true across the board for individuals and groups in any domain. That is, rapid progress toward significant goals depends upon high cognitive focus, excellent social-emotional engagement, high positive expectations, dynamic adjustments during interactions, and repeated encounters with challenges well beyond current performance. Most children *never* reach high art skills because they *never* encounter such a package or mix of conditions.

# Chapter 36

# INTEGRATION, REMINDERS, AND REFLECTIONS—MORE ON THE "TRICKINESS" WITHIN DYNAMIC TRICKY MIXING

I t is appropriate at this point to expand on one key earlier observation: both understanding the current dynamic patterns and predicting future patterns turns out to be *prone to error, complex,* and *elusive*—in short: *tricky.*

Despite all those aspects of trickiness, it is reassuring to look back at the whole range of previous examples of situations where tremendous advances did occur *if complexity-respecting new mixes were created, monitored, and adjusted.*

Another aspect of dynamic mixes is that many times there will be *fragility* for ongoing mixes.

Imagine, first, a Kentucky Derby race. Suppose that a fabulous, snowballing series of highly successful, positive dynamic tricky mix training strategies have prepared both a horse named Dragonfly and its jockey. As the race unfolds, Dragonfly has taken a narrow lead with just four hundred meters to the finish. A win seems assured. Then, one small change in the mix of conditions occurs, revealing the fragile dynamics. A competing jockey maneuvers illegally so that his horse slightly nudges and distracts Dragonfly, which leads to a brief stumble, loss of the lead, and then loss of the race itself.

Given such risks that a very small or brief intrusion into an ongoing positive mix may lead to disaster, below are some varied suggestions, which may help deflect and defend negative opponents and conditions. For the moment, we note here that in the case of leaders who are persistent threats to their communities, it is essential to understand the ongoing dynamics and find doable and affordable positive, creative, collaborative intrusions which succeed in disrupting harmful and illegitimate powers.

## Stuck "Doggies" Expedition: Identify New Stuck "Doggies" and Approach Re-Mixing with Optimism

Harking back to the earliest episodes in this book, think in terms of *stuck* "doggies." Where in your personal projects or broader social projects do you sense that a lack of progress has been accepted as unchangeable; that being *stuck* is both the obvious present and the obvious future?

A fantastic empirical base in the prior chapters has been created for, instead, believing that *if* active re-mixing, experimentation, and monitoring occur, then pathways which move beyond *stuck*-ness will be discoverable.

We have seen literally *stuck* Labrador retrievers rally their bodies to scamper, run, and climb stairs again after all hope had appeared to vanish. Ditto for creative, active re-mixing for humans who needed some new version of physical rehab. Ditto for crucial advances in organizations. And throughout the episodes among all chapters run the same two threads:

> The scope of possible advances will be underestimated when the range of attempted dynamic tricky re-mixing has been limited.
>
> Even a modest increase in understanding the dynamic complexities of situations and challenges opens the way to powerful, new tricky mix strategies.

## Construct Tricky Dynamic Support Structures to Ensure Smooth Performance Despite Predictable Future Pitfalls

Put some "training wheels" and "spotters" along the paths of your new project.

You will be monitoring closely because you know how tricky dynamics can be despite an overall excellent new mix or re-mix. Then you will be able to rush in just the right new "nudges" to the process when progress slows down.

In our work with new reading software for children, our monitoring showed that a few classrooms had reported that the software discs were duds. We replaced the discs and still they reported that the software would not launch. The only nudge we needed to solve this dilemma was to make sure that teachers followed a simple caution: no magnets on or near the discs!

## Under the Umbrella of Dynamic Re-Mixing, Utilize a Broad Set of "Re" Strategies: Regroup, Reexamine, Reframe, Recharge, and Re-Pivot

Dramatic improvements in an organization's effectiveness have been achieved by regrouping to utilize available resources in new mixes. For example, in schools, there are often mismatches between an individual student and the sole teacher assigned to their class. Regrouping will experimentally pair this student with other teachers for some hours every week, together with monitoring of social dynamics and academic progress. In this manner, new mixes are created, and those that are working well will be continued and then further improved. No new staff or materials are required in such re-mixes.

A related example is the typical American grade school, where there is a lack of highly skilled artists on the teaching staff, paired with a lack of adequate income-generating activities for many skilled artists in the local community. Research from my own lab demonstrated that even an hour or two a week can lead to explosive gains in a child's interest and abilities in making art—all that is required is a re-mix that provides enjoyable interactions between skilled artists and children which carry the dynamic combination of high challenges, positive emotion, and self-chosen experiments in producing art. Therefore, in the future, a feasible extension would be to reframe children's art education so that schools include highly skilled adult artists as the nucleus of child/artist interactions and create a re-mix of art activities that then leads to all children experiencing significant leaps in their art-making and their art-making enthusiasm.

## Otherwise Powerful and Positive Tricky Mixes Will Falter When Encountering Organized Opponents with Ulterior Motives: Be Prepared to Deploy High Intensity and Dramatic Pivots

Here is a summary from author Lee McIntyre of concerns in this domain:

"Many scientists have found it incredible in recent years that their conclusions about empirical topics are being questioned by those who feel free to disagree with them based on nothing more than gut instinct and ideology. This is irrational and dangerous. The nihilism about evolution, climate change, and vaccines has been stirred up in recent years by those who have an economic, religious, or political interest in contradicting certain scientific findings. Rather than merely wishing that particular scientific results weren't true, these groups have resorted to a public relations campaign that has made great strides in undermining the public's understanding of and respect for science. In part, this strategy has consisted of attempts to challenge the science by funding and promoting questionable research—which is almost never subject to peer review—in order to flood news outlets with the appearance of scientific controversy where there is none. The result has been a dangerously successful effort to subvert the credibility of science."

We may label these kinds of active misinformation campaigns, whether in scientific research or elsewhere, as a "fire" to be stopped. Rather than fight fire only with fire—that is, just with insults and contrary misinformation—it is often more effective to *fight fire with water*. In this latter approach, dynamic tricky counter-mixes are created which douse the false, negative information "flames" with streams of public shaming of the negative human sources and employ the many smothering layers of humor, clarification, boycotting of relevant commercial products, and easy access to accurate sources of relevant information. All these components should be mixed and re-mixed with highly engaging, persuasive spokespersons and active experimentation of varied pathways to positive change.

## Combatting Greedy Environmental Tyrants

Worldwide, part of the dangerous assault on our overall environment rests squarely on rogue, profit-first, powerful corporate and political leaders. We must come up with new extensive international dynamic tricky mix initiatives which are effective counters to such tyrants. Our air, oceans, rivers, soil, food, biodiversity, and climate are all in decline. More than that, the horrifying array of toxins and plastics which proliferate for profit and remain unregulated do not impact our world as single agents. Instead, they create together huge negative interacting and converging clusters and snowballs.

In consequence, bold and broad efforts for saving our health and our planet are needed. Among the efforts we should undertake are reform of all major regulatory agencies to free them from corruption and open them to full use of science and collaborative creativity. They should freely and fully research toxin impacts and then act aggressively to regulate as indicated. The EPA, FDA, and WHO currently are feeble and highly politically restrained—all need radical restructuring and expansion along these lines.

For all these areas of planetary significance, we must go way beyond just playing "defense." Suppose, for example, that one is concerned about endangered species and the rising rates of extinction. New dynamic tricky mixes should not be aimed at just slowing such rates. Instead, find complex and powerful ways of actually *increasing* the abundance of wild creatures. Restore planet Earth to the rich diversity of two hundred or so years ago. What we have covered on macaw reserves and related projects should help inspire a cascade of breakthroughs.

## Seek Activities and Contexts Where Surprising Events Are Likely to Emerge

In the course of preparing for this book and writing all the chapters, I have been delighted by the memories that this process has activated. I have come to realize and appreciate anew the ways in which fortunate experiences as a child and young adult have provided an active and supportive matrix for trying to achieve later breakthroughs.

In my childhood days in Kansas there were truly surprising events which unfolded from some hot, ordinary, and even boring activities. From those days I will now share a riddle—interpret the following two phrases:

Red and black, friend of Jack. Red and yellow kill a fellow.

So, my reader, what could these two phrases be describing? Take a moment and see if any reasonable hypothesis occurs to you.

My experience with these phrases related to an activity I frequently undertook with my older brother Craig. We would take a few cloth bags and venture onto Kansas hillsides where there were many rocks scattered around. We would systematically turn over rock after rock. Sometimes we would find so few items of interest that—particularly on a hot, sunny afternoon—a bit of boredom would set in.

However, usually, marvelous surprises eventually turned up and made the whole strenuous effort entirely worthwhile. On some days we indeed found "a friend of Jack." This was a brightly colored, beautiful snake with bands of red, black, and yellow running all along the length of the body. He was a king snake, so named because the beautiful color pattern was as bright as a coral snake but also distinctly different. On the poisonous coral snake, red and yellow bands were adjacent to each other and, thus, "red and yellow kill a fellow." In Kansas pastures we did not need to fear coral snakes because they did not live there. So we would turn over a rock, hoping for a surprise, and if we saw those marvelous colored bands, we could quickly grab the snake and not fear a dangerous bite would ensue. Our cloth bags would carry our treasures, many kinds of snakes and lizards, back home where we could amaze more friends.

Most folks in Kansas did not go out searching for these particular kinds of reptilian surprises. The beauty of many kinds of snakes was simply unknown to them because they had never been in a context where they would meet the snakes. Another beautiful species of king snake which we loved to find was called the speckled king snake. Their pattern was one of gold speckles sprinkled all along a dark black background. Here we can make a comparison to fishing. Both in my own childhood and in my daughter Leilani's childhood—as I related earlier in this book—suspense built up when you baited a line and then launched a floating bobber that at any instant might jerk or bobble and signal that there's something holding on to your bait. No matter how many times you've gone through this fishing exercise, you just cannot know what surprises may emerge from beneath the surface of the water. In a way, we may say that my brother, myself, and our friends, trekking across the pasture and overturning stones, were acting as "fishers" for snakes.

So when we enter contacts and activities where surprising events are likely to emerge but we just don't know what the surprises will be or when they will pop up, *then* small breakthroughs will occur. Importantly, these end up not only being enjoyable at the time, but they also feed into and nourish later new mixes of curiosity, adventure, and experimentation.

Another form of experimentation occurs in biology labs. The lab I want to tell you about now is the biology lab from my final semester as an undergraduate at Harvard University. The professors decided to help us organize experiments on frog eggs. These were exciting experiments because they related to groundbreaking new experiments in France on how genes are activated to achieve growth.

What we did was essentially cheat on nature. Without fertilizing a frog egg, we played tricks on that egg. For example, we would give it a brief dose of radiation. When the tricky mix of stimulation we provided was just right, a remarkable cascade of events was released. Relying on only half of the genes which usually support the gradual evolution from a single cell to a tadpole and then a full frog, I was able to experimentally cause a single egg to evolve into a large tadpole swimming around an aquarium tank. This tadpole swam and ate and behaved amazingly well despite missing half the expected genetic material.

Half the genes! This episode is a reminder that fits other episodes in this book: even with limited resources, breakthrough changes can be created with complexity-respecting new dynamic tricky mixes.

The only experiments I've ever run in biology were conducted in that single undergraduate lab. Nevertheless, looking back now I realize that out of the dramatic surprises through our own lab experiments there were lasting effects on me. The lab experiences became a powerful thread of my own deep interest in conducting other experiments that create clear-cut, causal outcomes. As you have already seen in this book, experiments in understanding developmental processes in children and creating new, innovative facilitating experiences for children has been part of my experimentation. But beyond that, I have gradually evolved a deep interest in how planful, complexity-respecting, and innovative experimentation is satisfying and productive in our personal and social lives as well as in the conduct of rigorous science.

## One Size Never Fits All

This notion is one that we usually fervently follow when we have decisions to make as consumers. Each of us wants the item we believe best fits our needs, preferences, and budget. So we research and seek the "best fit" for us in cell phones, running shoes, music, cars, energy bars, craft beers, and so on.

Paradoxically, though, our critical assessment powers are often laid aside in favor of passive acceptance of how major organizations are impacting us and our children.

Thus, for the most part, we let the "experts" proceed without challenge or consultation with us on their simplistic, one-size procedures for schools, ocean and wilderness management, economic stimulants, medical treatment plans, transportation, controls on pollution and other toxins, and climate change issues.

However, with dynamic tricky mix awareness, we should be prepared for active experimentation, monitoring, and adjustments to fight in all situations against the tyranny of one-sizing.

## Quick Take: Why Aren't Siblings as Similar as Many Expected Them to Be?

Take a situation where a family remains stable in terms of two parents remaining in a marriage until three siblings are raised to the age of eighteen. Doesn't that mean that the "environment" for each child is highly similar?

And if that is true, shouldn't we expect that all siblings in a stable family will show strong similarities in behaviors?

The problem you, dear reader, will probably spot at once—given all the related examples discussed so far—is one of over-simplification. Rather than each successive child encountering a simple "family environment,", what matters for a child's development is the dynamically created patterns of interactions they experience both with other siblings and with the parents. The complexity of each person involved changes over time in what they bring to the interactions. Further, countless contextual variations will all add up to significantly different environments for each child. What holds true for human children also holds true for the dynamics of experiences within a family of other creatures, including the tigers below, with multiple siblings.

*Two of Three Triplet Siberian Tiger Cubs/Siblings Strolling Off to Play.*

Conceptually, we may say the same about other named environmental contexts such as a Head Start center or other preschool. What happens under any labeled preschool with the same overall stated plans will be substantially different for different children. Accordingly, rather than expect uniform impacts or developmental outcomes, we should expect high variations. Rather than a simply described common environment, there are emerging and complex interplays of multiple sorts. So, to echo what we just observed for family settings, in preschool settings what matters to a child's development is the dynamically created patterns of interactions they experience both with other students and with the teacher and aides.

This is why we reported earlier in the book such startling contrasts in how much a four-year-old with no English skills can learn in a preschool year—in just that year some kids become as fluent in English as the kids who have been learning English all their lives while others don't get beyond a few scattered English words.

## Defying the "Gurus of Probability" Is a Path to Negative Cascades and "Black Swan" Events

Sometimes it is nearly impossible to calculate the probability of relevant conditions for a tricky mix. On the other hand, when we have good reason to trust informed "guru" estimators of probabilities, it is folly to ignore these inconvenient "gurus of probability" and substitute our own poorly grounded estimates instead.

Many a yachtsman has learned this lesson the hard way. Take the situation where in late summer to early autumn there is a desire to move one's yacht from the east coast of the US across the Atlantic to the coast of Greece. Hmm. What about hurricanes moving in? Estimates of good weeks for the Atlantic crossing in order to avoid hurricanes will be available. But when those are ignored and personal preferences are injected into the mix, struggles, boat damage, and risks to life all cascade toward potential disasters.

The book and movie entitled *Perfect Storm* capture negative events mixing dynamically toward disaster during swordfish fishing far offshore. Three different storm patterns were predicted to converge in a highly unusual, terribly strong storm—an extremely rare "black swan." Captains and crews defied the repeated observations of converging negative conditions. By making fishing their priority, along with a sense of being unusually lucky, they created a negative snowball which included being in the most treacherous waves and winds, suffering severe boat damage, incurring injuries to crew, and positioning themselves beyond any realistic chance of rescue. In the end, most of the crew suffered cruelly and lost their lives.

Another glaring "black swan" for the year of 2020 was the global pandemic caused by COVID-19. All indications are that the scope of this tragic pandemic could have been dramatically lessened if the available "demons of probability" had not been repeatedly defied.

From prior viral outbreaks, public health officials and citizens alike had available many proven strategies which could have lowered the probability of COVID-19 spreading and the serious health complications and deaths that resulted. The "demons of probability" concerning viruses were speaking loudly, trying to get people's serious attention. What should have been put in place was a bold, highly aggressive tricky mix of all the potent strategies available. Again, it must be stressed that a fully convergent, dynamic, multi-action mix could have been and should have been rapidly created.

Instead, in most states in the US and the majority of countries worldwide, defiance of appropriate actions was widespread. Priorities were placed on profits, personal appearances not being "marred" by masks, enjoyment of travel despite risks, a sense of being too lucky to be infected, and insider government contacts allowing exploitation of the crisis for personal benefits. Further worsening the accelerating crisis were frequent attacks on anyone and any strategy which intruded on personal choices, movements, freedom, and on the simple comforts found in usual patterns of life.

## Dynamic Puzzle #12: Very Lopsided Score

Dynamic Puzzle #12 posed the question: why would a crazy difference in score for a "Super Bowl of the future" be as big as two-hundred-ten to seven?

Well, the "Super Bowl" we have been playing for the years 2020 and 2021 is the global COVID-19 pandemic.

The puzzle's answer is that two modernized, technological societies have scored radically differently. One, through horribly poor mixing of strategies and tragic denials, is doing thirty times worse than the other in terms of both total cases per 100,000 and total deaths per 100,000. We in the USA are the losers who scored seven points in this metaphor. South Korea is the opposing country with excellent dynamic and complex tricky mixes and 1/30 of the negative outcomes seen in the USA.

## Recap and Extrapolation: How Early Paleolithic Breakthroughs Show the Way Forward to Modern New Tricky Mix Breakthroughs

We will now take a look at multiple natural shells from a collection of found symbolic objects which I personally collected. They connect much about our past and our potential futures.

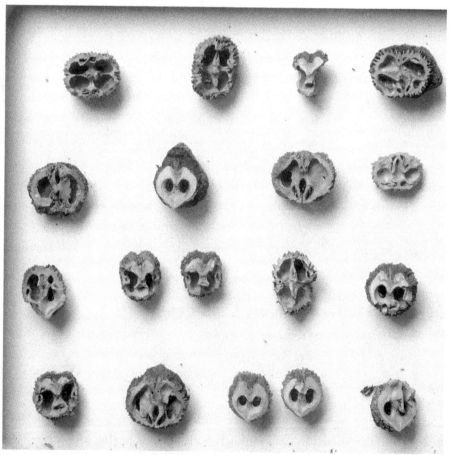

*Walnut Shells Found in Pennsylvania (or Anywhere in Prehistory)*
*Could Easily Symbolize Faces of Primates, Owls, or Other Animals.*

In Chapter 30, we saw that long ago—200,000 to two million years ago—our prehistoric hominin ancestors made a series of crucial breakthroughs by finding, reinterpreting, and sharing familiar natural objects. Without any initial change in their brains, a series of remarkable cultural changes were accomplished.

> The very first symbols were natural objects such as the walnut-shell symbols I called to your attention.
>
> Certain clans began collecting objects and treating seashells, rocks, nuts, and more as symbols.
>
> The very first art objects were these same collected, transported, and socially shared natural objects.
>
> A partial set of features—e.g., only a face and torso—can work today or two million years ago to communicate a symbolic reference to a conspecific.
>
> The very first tools were also natural objects. Some were pieces of bones or branches sharp enough to use for digging ground or scraping hides.
>
> The very first weapons also were natural objects. Among these were stones round enough and light enough to be easily hand-held and hurled at prey or in aggression toward others.

One key lesson for our seeking modern breakthroughs is this: new looks at objects and situations can support radical reinterpretations of what is possible. New dynamic mixes of meanings, attitudes, experiments, and communications can trigger initial breakthroughs followed in many instances by cascades of improving mixes and positive outcomes.

For our early Paleolithic ancestors, we gave a fairly full account in Chapter 30. In a nutshell (haha), the first levels of breakthroughs were that clans and individuals who led the way in new, profound uses of found objects gained survival advantages—more food, fewer deaths, and more offspring. Over many generations, the gene pools of the good finders and sharers moved toward more complex brains and even better skills in symbols, art, tools, and weapons. This cascade of changes in brains and co-evolving culture was set off initially neither by sudden mutations nor by shaping objects with tools. Instead, simple but powerful re-mixing of how collected and shared natural objects were interpreted and used triggered remarkable co-evolution. In short, these ancestors *found* their way to new levels of sophistication in culture and in brains.

*Round Stones for Throwing as Weapons Were Collected in Africa by Early Hominins 1.8 Million to 0.5 Million Years Ago.*

We can also recap our revised understanding of early hominin co-evolution in yet another fashion. The old story in anthropology and archaeology was that artifacts which were shaped/crafted/created are the prize artifacts for museum displays and theoretical emphases. Further, it was held that advances in sophistication of these artifacts was the key driver for advances in brain sophistication.

In contrast, the breakthrough, evidence-based new narrative is that advances in *collecting and sharing* significant natural objects drove survival value and advances in brain sophistication. These more advanced brains were then applied to reflection on collected tools, art, symbols, and weapons. Such reflection, in turn, triggered wave after wave of planning and creating/improving cultural artifacts. Cascades of positive and improving dynamic mixes then supported co-evolution of richer and richer communication, art-making, tool-making, weapon-making, and transport-making culturally along with the dynamic convergence of more sophisticated brains, more cooperative planning, and accelerated cross-clan contacts and sharing.

## Complexities of Symbolic Communication Revealed in Just Two Images

I will now show you an image of a horse head created with a digital camera. Depending upon context, a viewer may interpret this as an image only of a head, as a symbol of an entire horse, as a representation of a particular Kentucky Derby winner, and so on. Likewise, we have seen that prehistoric images of horses 30,000 or more years ago were created by artists in some clans on cave walls and ceilings, in statuettes, and in carvings on weapons. The artistic "tool kits" for these ancient artists are clearly superior to the "tool kits" of the vast majority of modern children at ages six to eight despite the seeming advantage of these modern children having already viewed millions of symbolic images in a wide variety of media.

The second image shown here is one that has deeply puzzled many archaeologists and anthropologists. The artist who created it made an overall pattern from stone that already had natural features

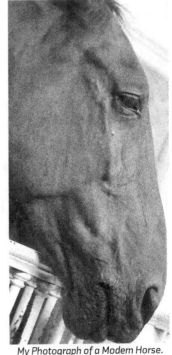

*My Photograph of a Modern Horse.*

with symbolic potential—for example, for portraying an eye and snout. The outline shape is that of a hand axe tool. However, this art piece is the size of a modern take-out pizza, and so it is too large and awkward to have *any* use as a tool. 400,000 years ago at what is now Swanscombe in the UK, considerable time and effort was invested to make a striking and beautiful *symbol*. In that region, horses were already common sights and of significance to humans. We can understand this singular art object as just one more product supported by dynamic tricky mix conditions for symbolic behavior—which, as we have shown, were present in some clans not only at 400,000 years ago, but long before that.

We cannot say for sure the precise meaning of this Swanscombe artifact.

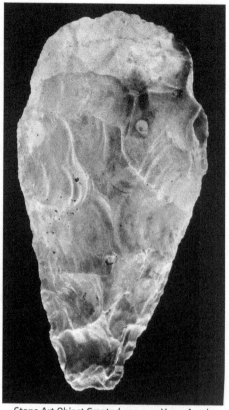

*Stone Art Object Created 400,000 Years Ago in Present-Day England—Often Mislabeled as a "Hand axe."*

However, we can say it is unquestionably symbolic. In addition, we know that it would have served—if the clan had reached consensus—powerfully as a symbol for horses, for animals in general, for animal spirits, as a symbol for the clan itself, or for a bond between humans and animals.

## Update on Predicted Finds at More Prehistoric Sites with Collected, Natural, Symbolic Objects

In a series of papers of mine beginning in 2015—and in earlier sections of this book—I explicitly predicted that more evidence of the Paleolithic collection of symbolic objects would be discovered. Now, in April 2021, as this book moves from draft toward publication, there is striking confirmation.

Here is Willoughby's capsule account of the significance of twenty-two natural crystals dating to 105,000 years ago. All the crystal objects were small, easy to fit in a hand, and were transported multiple miles to a Kalahari rock-shelter site in Southern Africa.

"Wilkins and colleagues systematically examined all the reasons why these pieces might be present there. After dismissing all possible natural explanations, in the same way that forensic researchers eliminate alternative scenarios during an investigation, they were left with only one conclusion: people intentionally collected such non-utilitarian objects. Their deposition presumably had some symbolic purpose."

Precisely as predicted! As I argued earlier in detail, we can expect many more such discoveries of the collection of natural symbolic objects all across the period of 100,000 to more than one million years ago.

Clans who were more adept at collecting such natural symbolic objects and agreeing upon their meaning within the clan would have had many advantages. Communication precision would rise, which then would lead to more success in finding food and avoiding danger, thus raising survival rates.

Below you will see images of these crystal symbols.

*Some of the Crystal Symbols in the Kalahari Dating to 105,000 Years Ago.*

Suppose for the moment that a clan chose to use specific, newly collected crystals as symbols for these referents: antelope, many/few, close/far, melon, fish, dangerous snake, men, women, children, and morning/evening. Then the clan might, for the first time, mix together explicit directions through the symbols for new plans. Laying out different combinations would then yield visual "utterances" of these kinds of messages: "Men morning find/kill many snakes. Women children evening close-by find melons. Men evening go-far get many fish and antelope."

It is essential, again, to stress that today, as well as 105,000 or more years ago, the meaning of any single visual symbol or any combination of symbols lies, not in the objects themselves, but rather in the socially negotiated meaning agreed upon within the culture. For the clans sharing the Kalahari crystal symbols, it is as of yet unclear how many of the meanings were along the lines of the examples above and how many may have touched on spiritual and/or mythic meanings.

As argued in more detail earlier, expansion of symbolic behaviors would facilitate survival value within a symbol-user's lifespan and even more so across successive generations. Such advances in the precision of communication and resulting increases in efficiency would all dynamically feed rapid snowballs of the co-evolution of more complex culture and more sophisticated brains.

## Two Key Reminders About the Dynamic Tricky Mixing and Re-Mixing Approach to Life: (1) Move Beyond Simplistic Thinking, and (2) Conduct Diverse Explorations and Then Actively Compare and Contrast Experiences

No doubt you will recognize how many examples across all chapters illustrate those themes. You may find it productive to see what happens to your idea generation if you revisit different chapters and actively seek new comparisons between the persons, situations, problems, and particular dynamic re-mixing patterns which greatly improved progress toward goals.

## More on The Dynamics of Rare Events

Stories of success most often cite long periods of persistent, gradual efforts despite many difficult circumstances and re-mixing adjustments along the way.

And yet, sometimes lives change in an instant. Within our Dynamic Systems framework, you have already seen that we label moments of immediate transformation as "rare events." Such events perturb and rearrange many aspects of our existing self-concepts, strategies, goals, beliefs, and focus.

Case in point: a single painting, photograph, or live performance will sometimes have the power to inspire a lifelong career path, trigger vast exploration and discovery, or even save the life of a depressed or confused person.

## Taking a Plunge

I had an experience one summer in an art studio. After decades of absorbing experiences viewing art in major and minor galleries and museums around the world, I decided to enroll in a basic painting course.

It was the first time in my life I had ever approached a blank canvas to attempt a painting of any kind. My instructor was a talented artist with a rich portfolio of abstract paintings right in his office. With his wonderful mentorship, I was *flowing* along with the canvas and the paint even in that initial painting. I now understood at a whole new level what countless artists through history had experienced in their work. Picasso's vivid description of how a relationship develops and deepens between the artist and the emerging stages of a painting. Painting is not my central career, but painting has expanded and enriched my explorations, collaborations, problem-solving, and creativity in many realms. Below I share a couple of my paintings with you. I invite you to explore any comparisons/relationships between these images and other images in this book.

"*Deep Blue Sea 1*" Acrylic on canvas.

Some say each pattern in nature connects to all other patterns in nature. What new images come to mind for you when viewing this painting?

*"Floating, Emerging Forms" Acrylic on canvas.*

You are invited to embrace ambiguity and connectedness. Do you see dragonflies in flight? Or the nucleus of a cell? Creatures floating somewhere on water? Or . . . what?

## Godspeed to You

You have explored a great diversity of situations in this book. As you move into the future pathways of your life, I wish you discovery, excitement, remarkable collaborations, creativity, emotional connections, and successes.

Please keep in mind what has emerged again and again. Namely, that each experience and pattern in nature connects to and illuminates all other experiences and patterns in nature.

Accordingly, if you approach your new experiences with a spirit of open exploration, a respect for complexity, curiosity about the perspectives and pathways of others, and an awareness of the power of dynamic tricky mix experimenting, you definitely will enhance your prospects for new breakthroughs.

Both these young raccoons are 60 feet up in a willow tree—an amazing way to evade predators. The typically-colored raccoon is up high as one of three surprisingly arboreal triplets, and the white raccoon is an Albino. Even though odds of spotting such an Albino are estimated at less than 1 in a million—less probable than your getting struck by lightning—we were delighted to see this one at our PA farm. No doubt our success was helped by our many eyes-wide-open hikes we take each week.

The typically-colored raccoon is one of three surprisingly arboreal triplets and the white raccoon is an albino. Even though the odds of spotting such an albino are estimated at less than one in a million—less probable than your getting struck by lightning—we were delighted to see this one at our Pennsylvania farm. No doubt our success was helped by the many hikes we take each week with our eyes wide open.

*Box turtles like this often have long lives—one-hundred-eighty years or more.*

By design, they have advantages of a great defense (their shell), low energy demands, and a strong and resilient heart. Look closely at this beautiful turtle shell. Maybe it resembles ancient Mayan art and writing or an Aborigine warrior's shield, or maybe it even represents multiple converging conditions of a dynamic tricky mix?

# ABOUT THE AUTHOR

*Author with Background of His Large Dynamic Process Painting.*

Growing up on the plains of Kansas, Keith Nelson was fortunate in having multiple tutors into the natural world including his father, mother, older brother, and boyhood friends. In the end, a series of explorations over his childhood years led to familiarity with and appreciation of just about every living creature in the local ecologies. College at Harvard then led to a PhD in psychology from Yale and a continuing career in studying children and teaching psychology. Currently a professor of psychology at Penn State University, he's held previous professorships at Stanford University and the graduate faculty of the New School for Social Research.

Keith's work crosses boundaries between developmental psychology, educational psychology, communication disorders, linguistics, art education, dynamic systems, cognitive psychology, creativity and innovation, environmental toxins, and evolution. The heart of his research concerns the developmental processes behind significant advances by children in every area of communication. Spoken language, sign language, reading, writing, and art have all been examined both in typically developing children and in children with various delays. Basic research has been translated widely into new procedures for helping to accelerate developmental progress in children with autism, language delay, dyslexia, and deafness. Publications have included twelve volumes in the book series *Children's Language* as well as innumerable journal articles and book chapters.

The complexity of change processes in children and in the learning contexts we provide for them is seen by Dr. Nelson as highly similar to the dynamics of change in ecological systems, in animal and plant life cycles and speciation, in co-evolution of hominin species from 1.5 million to 100,000 years ago, and in the changeable relationships of people with the natural world. His favored theory for these complex and varied change processes is called Dynamic Tricky Mix Theory.

Looking to the natural world helps to create awareness of patterns that are relevant to anyone's daily life. These natural patterns serve also as a catalyst toward innovations in how we see children. Nature also inspires us to create new arrangements of conditions to better support their progress in critical areas of communication, thinking, and spirituality.

For Keith, nature patterns have also greatly influenced his work in abstract painting, sculpture, and photography. These art creations have appeared in galleries and arts festivals. As an adult, he has traveled widely and engaged in all sorts of activities that bring him close to nature—ocean sailing, hiking, meditating in natural settings, camping, kayaking, introducing children to nature experiences, nature photography, fishing, farm operation, political action to preserve wild spaces and a diversity of species, and just sitting and doing "nothing."

# ENDNOTES

L isted here are selected—rather than exhaustive—reference citations and links/pointers to related publications. All of these will be valuable in further locating informational sources.

## Preface Through Chapter 1

Nelson, Keith E, and J D Bonvillian. "Development of Sign Language Skills in Autistic, Retarded, and Aphasic Children," 1978.

## Chapter 2. Assorted Eye-Openers

Nelson, Keith E, J Welsh, S Camarata, Heimann, and T Tjus. "A Rare Event Transactional Dynamic Model of Tricky Mix Conditions Contributing to Language Acquisition and Varied Communicative Delays," 2011.

## Chapter 4. Brief Sketch of Dynamic Systems and Dynamic Tricky Mix Theory

Nelson, Keith E. "A Dynamic Tricky MIX Theoretical Perspective on Acquiring L1 and L2 Spoken Language, Sign Language, Art, Text, and Other Symbolic Systems," 2005.

Thelen, Esther, and Linda B. Smith. *A Dynamic Systems Approach to the Development of Cognition and Action.* The MIT Press, 1996.

## Chapter 6. Why the F**k Not?

Nelson, Keith E. "Dynamic Tricky Mix Theory Suggests Multiple Analyzed Pathways (MAPS) As an Intervention Approach for Children with Autism and Other Language Delays," 2001.

## Chapter 7. Shooting Sacred Cows & Re-Mixes That Astonish

Diener, Ed, and Robert Biswas-Diener. *Happiness: Unlocking the Mysteries of Psychological Wealth*. Blackwell, 2011.Uhl, Christopher, and Dana L. Stuchul. *Teaching as If Life Matters: The Promise of a New Education Culture*. John Hopkins University Press, 2011.

## Chapter 8. Classic Over-Simplification Errors

Dietert, Rodney. *The Human Superorganism: How the Microbiome Is Revolutionizing the Pursuit of a Healthy Life*. Dutton Books, 2016.

Nelson, Keith E. "Progress in Multiple Language Domains by Deaf Children and Hearing Children: Discussions with a Rare Event Transactional Model," 1998.

Nelson, Keith E, J A Welsh, E M Vance Trup, and M T Greenberg. "Language Delays of Impoverished Preschool Children in Relation to Early Academic and Emotion Recognition Skills," 2011.

Williams, Donna. *Nobody Nowhere: The Extraordinary Autobiography of an Autistic*. Perennial, 2002. Uhl, Christopher, and Dana L. Stuchul. Teaching as If Life Matters: The Promise of a New Education Culture. John Hopkins University Press, 2011.

## Chapter 9. Surprising Recoveries and Life Path Shifts

Doidge, Norman. *The Brain's Way of Healing: Remarkable Discoveries and Recoveries from the Frontiers of Neuroplasticity*. Penguin Group US, 2015.

van der Kolk, Bessel. *The Body Keeps the Score: Brain, Mind and Body in the Healing of Trauma*. Penguin Books, 2015.

## Chapter 10. Extremely Rapid Changes & The Rare Event Principle

Nelson, Keith E, and J D Bonvillian. *Children's Language: Volume 3*, 1982.

Nelson, Keith E. *Children's Language: Volume 6*, 1987.

Nelson, Keith E. "Methods for Stimulating and Measuring Lexical and Syntactic Advances: Why Fiffins and Lobsters Can Tag along with Other Recast Friends," 2000.

Sacks, Oliver W. *An Anthropologist on Mars*. Alfred A. Knopf Canada, 1995.

Sacks, Oliver. *Musicophilia*. Picador, 2018.

## Chapter 11. Quantum Shifts: Quantitative Underpinnings of Major Qualitative Shifts in System Performance

Collins, James C. *Good to Great: Why Some Companies Make the Leap . . . and Others Don't*. Collins, 2009.

Csikszentmihalyi, Mihaly. *Flow: The Psychology of Optimal Experience*. Harper and Row, 2009.

Sharma, Monica. *Radical Transformational Leadership: Strategic Action for Change Agents*. North Atlantic Books, 2017.

## Chapter 13. We Met Once Twenty Years Ago, but Let Me Show You I Remember Your Shoes, Buttons, Food Order & Words

Nelson, Keith E, and M E Arkenberg. "Setting the Stage: Developmental Cognitive and Interactional Underpinnings of Language and Reading from a Dynamic Systems Perspective," 2008.

Mody, M, and E R Silliman. *Brain, Behavior, and Learning in Language and Reading Disorders*, 2008.

LePort, Aurora, Aaron Mattfeld, Heather Dickinson-Anson, James Fallon, Craig Stark, Frithjof Kruggel, Larry Cahill, and James McGaugh. "Behavioral and Neuroanatomical Investigation of Highly Superior Autobiographical Memory (HSAM)," 2012.

## Chapter 14. Late-Talking Children: Breakthroughs and Slain Sacred Cows

Nelson, Keith E, and S M Camarata. "Conversational Recast Intervention with Preschool and Older Children," 2006.

Nelson, Keith E, M T Clarke, and G Soto. "Language Learning, Recasts, and Interaction Involving AAC: Background and Potential for Intervention," 2017.

Cleave, Patricia L, Stephanie D Becker, Maura K Curran, Amanda J Owen Van Horne, and Marc E Fey. "The Efficacy of Recasts in Language Intervention: a Systematic Review and Meta-Analysis," 2015.

Farrar, Michael J. *Discourse and the Acquisition of Grammatical Morphemes.* Cambridge University Press, 2009.

Nelson, Keith E. "Dynamic Tricky Mix Theory Suggests Multiple Analyzed Pathways as an Intervention Approach for Children with Autism and Other Delays," 2002.

## Chapter 15. The Surprising Dynamics of Long-Necked Giraffes and Fish with No Ribs

Carroll, Sean B. *Endless Forms Most Beautiful: the New Science of Evo Devo and the Making of the Animal Kingdom.* Quercus, 2012.

## Chapter 16. All Calm in the Dentist's Chair

Davidson, Richard J, and Antoine Lutz. "Buddha's Brain: Neuroplasticity and Meditation," 2008.

## Chapter 17. Seek Complex Changes in Self Rather Than Simple Knowledge or Skill Increases

Nelson, Keith E, M Lundalv, T Tjus, and M Heimann. "Animega: A New Mobile App for Facilitating Research and Facilitating Teaching Children in the Domains of Literacy, Language, and Creativity," 2019.

Bugental, James Frederick Thomas. "Psychotherapy and Process: The Fundamentals of an Existential-Humanistic Approach. Addison-Wesley," 1978.

Collins, B. "My Life," 2004.

Kelley, Tom, and David Kelley. *Creative Confidence: Unleashing the Creative Potential within Us All*. William Collins, 2015.

Matson, Jack V. *Innovate or Die!: A Personal Perspective on the Art of Innovation*. Project Management Institute, 2013.

Nelson, Keith E, and P M Prinz. "A Child Computer Teacher Interactive Method for Teaching Reading to Young Deaf Children," 1985.

Uhl, Christopher, and Dana L. Stuchul. *Teaching as If Life Matters: The Promise of a New Education Culture*. John Hopkins University Press, 2011.

## Chapter 18. Ocean Sailing

Bergreen, Laurence. *Over the Edge of the World: Magellan's Terrifying Circumnavigation of the Globe*. William Morrow, 2019.

Oliver, Mary. *New and Selected Poems Volume Two*. Beacon Press, 2007.

Oliver, Mary. *West Wind*. Houghton Mifflin, 1997.

## Chapter 19. Startling Contrasts: Mostly Stuck vs. Zooming Along

De Houwer, Annick. *Bilingual First Language Acquisition*. Multilingual Matters, 2009.

Gopnik, Alison. *The Gardener and the Carpenter*. Vintage, 2017.

Lepper, M, M Woolverton, D L Mumme, and J L Gurtner. "Motivational Techniques of Expert Human Tutors: Lessons for the Design of Computer-Based Tutors," 1993.

Newton, M G, L G Castonguay, T D Borkovec, A J Fisher, and S S Nordberg. "An Open Trial of Integrative Therapy for Generalized Anxiety Disorder," 2008.

Welsh, Janet A., Robert L. Nix, Clancy Blair, Karen L. Bierman, and Keith E. Nelson. "The Development of Cognitive Skills and Gains in Academic School Readiness for Children from Low-Income Families," (2010).

## Chapter 20. When Labels Become Sacred Cows

Nelson, Keith E, M Heimann, and T Tjus. "Theoretical and Applied Insights from Multimedia Facilitation of Communication Skills in Autistic, Deaf, Motorically Disabled, and Children with Other Disabilities," 1997.

Tjus, Tomas, Mikael Heimann, and Keith E. Nelson. "Gains in Literacy through the Use of a Specially Developed Multimedia Computer Strategy," 1998.

## Chapter 21. Draculas, Dunces, and Other Obstacles to Innovation, Change, and Breakthroughs

Cameron, Julia. *The Artist's Way: A Spiritual Path to Higher Creativity*. Jeremy P. Tarcher/Penguin, 2002.

Cameron, Julia. *The Sound of Paper: Starting from Scratch*. Jeremy P. Tarcher/Penguin, 2005.

Chödrön Pema. *When Things Fall Apart: Heart Advice for Difficult Times*. Thorsons Classics, 2017.

## Chapter 22. Snowballs

Heimann, M, T Tjus, M Lundalv, and Keith E Nelson. "Animega: A New Mobile App for Facilitating Research and Facilitating Teaching Children in the Domains of Literacy, Language, and Creativity," 2019.

Nelson, Keith E, S M Camarata, J A Welsh, M Heimann, and T Tjus. "A Rare Event Transactional Model of Tricky Mix Conditions Contributing to Language Acquisition and Varied Communicative Delays," 2011.

Schwartz, Evan. *Juice: the Creative Fuel That Drives World-Class Inventors*. Harvard Business School, 2004.

Weiner, Jonathan. *The Beak of the Finch: A Story of Evolution in Our Time*. Vintage Books, 2014.

## Chapter 23. Comparing the Present "Dynamic Tricky Mix" Account of Complex Change to Other Perspectives

Firestein, Stuart. *Failure: Why Science Is So Successful*. Oxford University Press, 2016.

Duckworth, Angela. *Grit: The Power of Passion and Perseverance*. Scribner, 2016.

Matson, Jack V. *Innovate or Die!: A Personal Perspective on the Art of Innovation.* Project Management Institute, 2013.

Pink, Daniel H. *Drive: the Surprising Truth about What Motivates Us.* Riverhead Hardcover, 2009.

Thelen, Esther, and Linda B. Smith. *A Dynamic Systems Approach to the Development of Cognition and Action.* The MIT Press, 1994.

Vaillant, George E. *Aging Well: Surprising Guideposts to a Happier Life from the Landmark Study of Adult Development.* Little, Brown Spark, 2002.

Vaillant, George E. *Triumphs of Experience: The Men of the Harvard Grant Study.* The Belknap Press of Harvard University Press, 2015.

## Chapter 24. Are There Instincts for the Most Complex Human Performances and Abilities?

Carroll, Sean B. *Endless Forms Most Beautiful: the New Science of Evo Devo.* W.W. Norton & Company, 2005.

Christiansen, Morten H. *Creating Language: Integrating Evolution, Acquisition, and Processing.* The MIT Press, 2018.

Elman, Jeffrey L., A Karmiloff-Smith, E Bates, M Johnson, D Parisi, and K Plunkett. *Rethinking Innateness: A Connectionist Perspective on Development.* MIT Press, 1997.

Levitin, Daniel J. *This Is Your Brain on Music: The Science of a Human Obsession.* Langara College, 2006.

MacWhinney, Brian. *The Emergence of Language.* Psychology Press, 1999.

Nelson, K.E. *Children's language.* Volumes 1–11. 1978–2011.

Nelson, Keith E. "Varied Domains of Development: A Tale of LAD, MAD, SAD, DAD, and RARE and Surprising Events in Our RELMS," 1991.

Nelson, Keith E, and M E Arkenberg. "Setting the Stage: Developmental Cognitive and Interactional Underpinnings of Language and Reading from a Dynamic Systems Perspective," 2008.

Nelson, Keith E., Aran Barlieb, Kiren Khan, Elisabeth M. Trup, Mikael Heimann, Tomas Tjus, Mary Rudner, and Jerker Ronnberg. "Working Memory, Processing Speed, and Executive Memory Contributions to Computer-Assisted Second Language Learning," 2012.

Whyte, Elisabeth M., and Keith E. Nelson. "Trajectories of Pragmatic and Non-literal Language Development in Children with Autism Spectrum Disorders." *Journal of Communication Disorders*, 2015.

## Chapter 25. Butterflies and Dragonflies

Nelson, Keith E, and A San Jose. "Increasing Children's Positive Connection To, Orientation Toward, and Knowledge of Nature Through Nature Camp Experiences," 2017.

Mlodinow, Leonard. *The Drunkard's Walk: How Randomness Rules Our Lives.* Pantheon Books, 2009.

Tallamy, Douglas W. *Bringing Nature Home: How You Can Sustain Wildlife with Native Plants.* Timber Press, 2016.

## Chapter 26. Perfect Dynamic Storm: How Simplicities Ruin Criminal Justice Reform in the USA and Abroad

Einat, Tomer, and Amela Einat. "Learning Disabilities and Delinquency: A Study of Israeli Prison Inmates," 2007.

Miller, Reuben Jonathan. *Halfway Home: Race, Punishment, and the Afterlife of Mass Incarceration.* Little, Brown and Company, 2021.

## Chapter 27. Creating Remarkable Dynamic Events

Childs, Craig Leland. *The Animal Dialogues: Uncommon Encounters in the Wild.* Black Bay/Little, Brown, 2009.

Heath, Dan, and Chip Heath. *The Power of Moments.* Random House UK, 2019.

## Chapter 28. Deal Breakers, Naysayers, Wreckers, and Other Opponents

Cameron, Julia. *The Artist's Way: A Spiritual Path to Higher Creativity.* Jeremy P. Tarcher/Penguin, 2002. Cameron, Julia. *The Sound of Paper: Starting from Scratch.* Jeremy P. Tarcher/Penguin, 2005.

## Chapter 29. Humor as a Catalyst for New Effective Mixes for Small but Annoying Problems as Well as Larger Issues

Feynman, Richard P. *"Surely You're Joking, Mr. Feynman!": Adventures of a Curious Character*. W.W. Norton, 1985.

Feynman, Richard P. *What Do You Care What Other People Think?: Further Adventures of a Curious Character*. W. W. Norton, 1988.

Magliozzi, Tom, and Ray Magliozzi. *Ask Click and Clack: Answers from Car Talk*. Chronicle Books, 2008.

Vaillant, George E. *Aging Well: Surprising Guideposts to a Happier Life from the Landmark Study of Adult Development*. Little, Brown Spark, 2002.

Vaillant, George E. *Triumphs of Experience: The Men of the Harvard Grant Study*. The Belknap Press of Harvard University Press, 2015.

## Chapter 30. Da Vinci's Children, Da Vinci's Ancestors

Bednarik, Robert G. "The Earliest Evidence of Palaeoart," 2003.

Curry, A. "Truth and Beauty: The Discovery of a 40,000-Year-Old Figurine Reignites Debate among Archaeologists about the Origin—and True Purpose—of Art," 2012.

Henshilwood, Christopher S., Francesco D'errico, Curtis W. Marean, Richard G. Milo, and Royden Yates. "An Early Bone Tool Industry from the Middle Stone Age at Blombos Cave, South Africa: Implications for the Origins of Modern Human Behaviour, Symbolism and Language," 2001.

Lorblanchet, Michel, Paul G. Bahn, and Pierre Soulages. *The First Artists in Search of the World's Oldest Art*. Thames&Hudson, 2017.

Nelson, Keith E, and M E Arkenberg. "Setting the Stage: Developmental Cognitive and Interactional Underpinnings of Language and Reading from a Dynamic Systems Perspective," 2008.

Nelson, Keith E, and K S Khan. "Refinements in Theories of Narrative Skill Acquisition by Children and Proposals for Theoretically Derived Effective Narrative Intervention Programs," 2018.

Nelson, Keith E. "Key Roles of Found Symbolic Objects in Hominin Physical and Cultural Evolution: The Found Symbol Hypothesis," 2016.

Nelson, Keith E. "A Dynamic Tricky Mix Theoretical Perspective on Acquiring L1 and L2 Spoken Language, Sign Language, Art, Text, and Other Symbolic Systems," 2005.

Nelson, Keith E, P L Craven, Y Z Xuan, and M Arkenberg. "Acquiring Art, Spoken Language, Sign Language, Text and Other Symbolic Systems: Developmental and Evolutionary Observations from a Dynamic Tricky Mix Theoretical Perspective," 2004.

Potts, Richard. "Olorgesailie: New Excavations and Findings in Early and Middle Pleistocene Contexts, Southern Kenya Rift Valley," 1989.

Potts, R. "Small Mid-Pleistocene Hominin Associated with East African Acheulean Technology," 2004.

Tomasello, Michael. *Origins of Human Communication*. MIT Press, 2008.

# Chapter 31. Tricky Mix Explorations Which Create New Retrieval Paths, New Sensory Experience Paths, and New Connections Between the Unconnected

Abram, David. *Becoming Animal: an Earthly Cosmology*. Vintage Books, 2011.

Matthiessen, Peter. *Nine-Headed Dragon River: Zen Journals, 1969-1982*. Shambhala, 1982.

Nelson, Keith E. *Children, Pelicans, Planets: Bobcat Magic*. Super Impact Images Press, 2010.

# Chapter 32. Positive and Negative Snowballs in Nature's Ecological Systems

Cribb, Julian. *Poisoned Planet: How Constant Exposure to Man-Made Chemicals Is Putting Your Life at Risk*. Allen & Unwin, 2015.

Friedman, Thomas L. *Thank You for Being Late: An Optimist's Guide to Thriving in the Age of Accelerations*. Penguin Books, 2017.

Goldsworthy, Andy. *Passage*. Harry N. Abrams, 2004.

Goldsworthy, Andy. *Enclosure*. Abrams, 2007.

Kahn, Peter H., and Stephen R. Kellert. *Children and Nature: Psychological, Sociological and Evolutionary Investigations*. MIT Press, 2002.

Heath, Dan, and Chip Heath. *The Power of Moments*. Random House UK, 2019.

Latour, Bruno. *Facing Gaia*. Polity Press, 2017.

Latour, Bruno. *Politics of Nature: How to Bring the Sciences into Democracy*. Harvard University Press, 2009.

Louv, Richard. *Vitamin N: the Essential Guide to a Nature-Rich Life*. Atlantic Books, 2017.

Louv, Richard. *Last Child in the Woods*. Algonquin Books od Chapel Hill, 2006.

Valdivia, Abel, Shaye Wolf, and Kieran Suckling. "Marine Mammals and Sea Turtles Listed under the U.S. Endangered Species Act Are Recovering," 2019.

Ryn, Sim Vander, and Stuart Cowan. *Ecological Design*. Island Press, 1996.

## Chapter 33. More Tricky Mix Moves in the Face of High Uncertainty and High Complexity

Sharma, Monica. *Radical Transformational Leadership: Strategic Action for Change Agents*. North Atlantic Books, 2017.

## Chapter 34. Living Our Lives as Tricky Mix Experimenters: Review, Part One

Raz, Guy. *How I Built This: the Unexpected Paths to Success from the World's Most Inspiring Entrepreneurs*. Houghton Mifflin Harcourt, 2020.

Ramachandran, V. S., and Sandra Blakeslee. *Phantoms in the Brain: Probing the Mysteries of the Human Mind*. Harper Perennial, 1998.

Ramachandran, V. S. *The Tell-Tale Brain: Unlocking the Mystery of Human Nature*. Windmill, 2012.

Sacks, Oliver W. *An Anthropologist on Mars*. Alfred A. Knopf Canada, 1995.

Whyte, Elisabeth M., and Keith E. Nelson. "Trajectories of Pragmatic and Nonliteral Language Development in Children with Autism Spectrum Disorders," 2015.

## Chapter 35. Living Our Lives as Tricky Mix Experimenters: Review, Part Two

Weiner, Jonathan. *The Beak of the Finch: A Story of Evolution in Our Time*. Vintage Books, 2014.

## Chapter 36. Integration, Reminders, and Reflections

Backwell, L R, and F d'Errico. "The First Use of Bone Tools: A Reappraisal of the Evidence from Olduvai Gorge, Tanzania," 2004.

Backwell, L R, and F d'Errico. "Additional Evidence on Early Hominid Bone Tools from Swartkrans," 2005.

d'Errico, F, L R Backwell, and L R Berger. "Bone Tool Use in Termite Foraging by Early Hominids and Its Impact on Understanding Early Hominid Behaviour," 2001.

Smitha, Elaine. *Screwing Mother Nature for Profit*. Watkins, 2011.

Walter, M, and M Trauth. "High Concentrations of Acheulean Handaxes from the Kariandusi, Gadeb and Olorgesailie Prehistoric Sites: Reworked Artifacts or in Situ Tool Factories?," 2011.

Wilkins, Jayne, Benjamin J. Schoville, Robyn Pickering, Luke Gliganic, Benjamin Collins, Kyle S. Brown, Jessica von der Meden, et al. "Innovative Homo Sapiens Behaviours 105,000 Years Ago in a Wetter Kalahari," 2021.

Willoughby, Pamela R. "Early Humans Far from the South African Coast Collected Unusual Objects," 2021.

Wilson, Andrew D., Qin Zhu, Lawrence Barham, Ian Stanistreet, and Geoffrey P. Bingham. "A Dynamical Analysis of the Suitability of Prehistoric Spheroids from the Cave of Hearths as Thrown Projectiles," 2016.

# A free ebook edition is available with the purchase of this book.

**To claim your free ebook edition:**

1. Visit MorganJamesBOGO.com
2. Sign your name CLEARLY in the space
3. Complete the form and submit a photo of the entire copyright page
4. You or your friend can download the ebook to your preferred device

Morgan James BOGO™

A **FREE** ebook edition is available for you or a friend with the purchase of this print book.

CLEARLY SIGN YOUR NAME ABOVE

**Instructions to claim your free ebook edition:**
1. Visit MorganJamesBOGO.com
2. Sign your name CLEARLY in the space above
3. Complete the form and submit a photo of this entire page
4. You or your friend can download the ebook to your preferred device

## Print & Digital Together Forever.

Snap a photo

Free ebook

Read anywhere

Printed in the USA
CPSIA information can be obtained
at www.ICGtesting.com
LVHW041210210924
791747LV00002B/81

9 781631 956690